STRUCTURAL BRIDGE ENGINEERING

Edited by **Shahiron Shahidan, Shahrul Niza Mokhatar, Mohd Haziman Wan Ibrahim, Norwati Jamaluddin, Zainorizuan Mohd Jaini** and **Noorwirdawati Ali**

Structural Bridge Engineering

http://dx.doi.org/10.5772/61683
Edited by Shahiron Shahidan, Shahrul Niza Mokhatar, Mohd Haziman Wan Ibrahim, Norwati Jamaluddin, Zainorizuan Mohd Jaini and Noorwirdawati Ali

Contributors

Ahmet Can Altunişik, Alcínia Zita Almeida Sampaio, Slawomir Karas, Andrea Mardegan, Alessandro Contin, Alessandro Lipari, George C. Manos, Nikolaos Simos, Evangelos Kozikopoulos

CBS, Edition **2017**

Published by InTech

Janeza Trdine 9, 51000 Rijeka, Croatia

Publishing Process Manager Romina Rovan

Technical Editor SPi Global

Cover Designer InTech Design Team

Additional hard copies can be obtained from orders@intechopen.com

Structural Bridge Engineering , Edited by Shahiron Shahidan, Shahrul Niza Mokhatar, Mohd Haziman Wan Ibrahim, Norwati Jamaluddin, Zainorizuan Mohd Jaini and Noorwirdawati Ali
p. cm.
Print ISBN 978-953-51-2688-1
Online ISBN 978-953-51-2689-8

INTECH

open science | open minds

Contents

Preface

Over the past decades, the society requires the creation of new bridges to ensure good and effective infrastructure. This is true for both effective urban and rural areas and for roads, railways and other transport modes including bicycles. Therefore, the field of bridge engineering is one of the most related fields to meet those needs. Bridge engineering consists of a lot of scientific concepts and includes a huge investment of money. It is very much needed to design any bridge economically and construct it with proper knowledge and maintenance.

Bridge engineering has developed in line with the new millennium. Although the general techniques of bridge design have been established for thousands of years, every bridge presents complicated factors that must be taken into consideration, such as the geology of the surrounding area, the amount of traffic, weather conditions and construction materials. Sometimes these factors are miscalculated, or something occurs that the bridge designers did not expect. New construction techniques and new materials are emerging. There are also new issues faced by the bridge-building industry relative to the research associated with these new techniques and materials. A detailed description of each chapter has been made as follows:

History of Sustainable Bridge Solutions

This chapter discusses the concerns of sustainable bridge development. The presented approach refers to elements of design, construction and maintenance of bridges with regard to their role in transport and social life—which have been present in bridge construction for a long time and can be easily incorporated into the concept of sustainable bridge construction.

Highway Bridge Traffic Loading

In this chapter, the traditional approaches and recent advances in highway bridge traffic loading are described. These are of great significance for the structural safety assessment of bridges. In fact, it is widely accepted that consideration of site-specific traffic features results in significant savings in maintenance operations.

3D and 4D Models Used in Bridge Design and Education

This chapter discusses 3D and 4D model analysis used in bridge design. The aesthetic analysis is an important issue that must be considered when designing a new bridge, especially when it is to be built in an urban area. In this context, the automatic generation of three-dimensional (3D) geometric models of the bridge under analysis, and the walk-around and aerial simulation allowed over it, may help bridge designers to evaluate its aesthetic concept and environmental impact. The bridge construction process can also be simulated, helping designers and builders to review the progress of the construction work in situ. The simulation of the construction activity made possible by the developed interactive 4D model helps bridge designers to analyse the whole construction process.

The Structural Performance of Stone Masonry Bridges

This chapter covers the performance of structural stone masonry bridges, which is examined by studying such structures located at the North-West area of Greece which have been declared as cultural heritage structures. The dynamic characteristics of four stone bridges, obtained by temporary in situ instrumentation, are presented together with the mechanical properties of their masonry constituents. The basic assumptions of relatively simple 3D numerical simulations of the dynamic response of such old stone bridges are discussed based on all selected information. The results of these numerical simulations are presented and compared with the measured response obtained from the in situ experimental campaigns.

Different Solutions for Dissipation of Seismic Energy on Multi-Span Bridges

This chapter deals with the application of various methods for the dissipation of seismic energy in order to adjust the response to seismic forces of an existing bridge with multiple spans.

Fluid Viscous Dampers and Shock Transmitters in the Realisation of Multi-Span Steel-Concrete Viaducts

The chapter discusses the methodology for the dissipation of seismic energy, designed for the construction of a steel-concrete viaduct in a variable orography land. The viaduct has a total length of 1,102 m and a typical span of 75 m, with piers of a maximum height of 65 m. The viaduct is subjected to a redesign step in order to adapt it to the requirements of Italian standard "D.M. 14/01/2008". The new design has reformulated the sequence of spans and consequently redefined the structures constituting the foundations, piles, steel girders, inferior bracings and especially the typology of bearings and seismic devices.

Nondestructive Testing Structural Bridge Identification

This final chapter presents the determination of bridge responses during their service life, which has gained great importance using non-destructive test methods with the change of aims, usage, environmental conditions, material deterioration by time and damages during certain dramatic events. In this chapter, non-destructive experimental measurement test results of bridges are presented for structural identification. The measurements are conducted under environmental excitations of pedestrian movement, traffic and wind-induced vibration, while the response signals are collected using uniaxial and triaxial sensitive seismic accelerometers.

To sum up, I would like to thank all authors for their dedication and willingness to contribute to each chapter found in this book. I am also grateful to my coeditors, Associate Professor Dr. Mohd Haziman Wan Ibrahim, Dr. Shahrul Niza Mokhatar, Dr. Norwati Jamaluddin, Dr. Noorwirdawati Ali and Dr. Zainorizuan Mohd Jaini, for their kind assistance in reviewing this book. All constructive criticisms and suggestions received have contributed immensely in the publication of this book.

Dr. Shahiron Shahidan
Universiti Tun Hussein Onn Malaysia,
Jahor, Malaysia

Chapter 1

History of Sustainable Bridge Solutions

Slawomir Karas

Additional information is available at the end of the chapter

http://dx.doi.org/10.5772/63461

Abstract

The chapter is a voice in the discussion concerning sustainable bridge development. Nowadays, the term has rather been abused, and therefore the presented approach refers to these elements of design, construction and maintenance of bridges—with regard to their role in transport and social life—which have been present in bridge construction for a long time and can be easily incorporated into the concept of sustainable bridge construction. Sustainable development, sustainable construction and so on are multidimensional. In the considered bridge construction area, looking at construction processes as interfering with the environment and which could and should be restricted is a new element. Nevertheless, other proven constructional solutions and technologies are characterised by their reliability. Assuming that the constructed bridges are to serve the next two or three generations of users, we can try to extrapolate current technical conditions on the next 30 or 60 years, i.e., up to three generations. We can do it if we know and are able to critically assess the history of bridge construction. Following this reasoning, the history in question is referred to in this paper, although rather subjectively and with the omission of numerous important personalities and technologies as well as instructive failures due to the publishing limitations.

Keywords: Bridges, History of Bridges, Aesthetics, Sustainability, Architecture

1. Introduction

Sustainable bridges is a term exemplifying the general idea of sustainable development. The concept is a result of the works of a UN commission—the World Commission on Environment and Development—conducted from December 1983 to December 1987 and concluded with the report *Our Common Future*. At present, several definitions of the basic concept and its specific component disciplines are in use. One of the general definitions emphasises preservation of natural environment by achieving the developmental objectives in a responsible way: *the present generations' responsibility to regenerate, maintain and improve planetary resources for use by future generations*. Nowadays, the idea has developed into specific disciplines, such as sustainable engineering, sustainable bridges [1] and sustainable design [2].

In the development of civilisation, an extensive and well-organised transportation system, i.e., roads, railway, bridges, air transport and maritime and inland navigation, safe both for people and environment, is of primary importance. The development and constant modernisation of road and rail infrastructure is connected with significant pollution emissions into the air, soil and water. Investments are accompanied by noise and traffic disruptions. These negative side effects have an equally strong impact both on people and environment. In general, they can result in a temporary or even permanent closing or changing of wildlife corridors, animal herd fragmentation, changing of nesting sites or habitats.

Roads and bridges, despite obvious differences, constitute a technically inseparable set. The name of the first and still functioning technical school, excellent *École nationale des ponts et chaussées* founded in 1747, is quite symptomatic here. Bridges as such belong to the elite area of civil engineering. This position stems from their civilisational, social, architectural as well as military role.

After the horrible experience of the first World War, the Briand-Kellogg treaty [3], renouncing war as an instrument of national policy, was concluded. History, including the most recent one, shows that military operations are still conducted on a different scale and so it happens that some are aimed at bridges. **Figure 1** shows the Hanoi bridge that was destroyed during the Vietnam War. Its crippled form is a dramatic monument to this war.

Figure 1. Hanoi Cầu Long Biên Bridge (1903) over the Red River—a war monument.

Sustainable bridge construction refers to the minimisation of harmful emissions during bridge construction. It is equally important, however, to design durable bridges, i.e., with minimum serviceability of 100 years as stipulated in the design standards [4]. The bridge longevity results from the structure maintenance therefore its design should take into consideration the ease of its future maintenance.

A design taking into account the future changes of traffic, both in terms of its volume and type, seems to be a great challenge. The existing methods of forecasting traffic changes cover periods from 5 to 10 years. Nobody can foresee what can happen in 50 years, [5]. Bridges have been built for thousands years and that is why it is easy to indicate the ones which have proved reliable according to various criteria. For this reason, they can be seen as sustainable bridges. The history of bridge construction referred to this paper is presented from such a perspective. In many cases, only the name of a bridge is mentioned, which results from its recognisability as well as an easy access to basic encyclopaedic information on the Internet.

2. Archaic period

The history of bridges can be told in many ways. It will always be a subjective representation, strongly affected by the author's attitude. This is also the case here.

The first large-scale bridge, recorded by Herodotus, was a structure for crossing the Dardanelles (Hellespont in the ancient times, the area of Çanakkale in Turkey at present), constructed by Greek constructor Mandrocles of Samos in 513 B.C. It was a ship pontoon bridge. Soon after, in 480 B.C., two other pontoon bridges were constructed by Xerxes' army, see *History of Herodotus* [6]. At the point of the bridge crossings, the Strait of the Dardanelles is 1.4 km wide. It should be noted that Herodotus' description of the bridges gives rise to doubts; therefore it can only be assessed in terms of likelihood.

In 55–53 B.C., during the wars with the Germanic peoples, Julius Caesar commissioned the construction of two wooden bridges over the river Rhine. At least two reconstructions of these bridges exist [7]. Another large-scale bridge was constructed during Emperor Trajan's war against Dacia in 105 A.D. by Greek architect Apollodorus of Damascus. Its remains, in the form of stone pillars, are to be found in a place with a poetic name: Turnu Severin. It was a wooden arch structure based on 20 massive supports where the Roman pozzolana cement was used. The bridge was immortalised in a relief on Trajan's column in the Foro di Traiano in Rome.

Figure 2. Arch-corbel system of Mycenaean bridges: (a) Bridge A; (b) Bridge B.

All of the aforementioned bridges do not exist anymore, but in Peloponnesus there are still two stone bridges constructed with so-called *Cyclopean boulders* to be found, known as Mycenaean Bridges. The period of the Mycenaean culture falls in the mid-Bronze Era from approx. 3000 B.C. to 1000 B.C. Hence, it can be inferred that the bridges are at least 3000 years old and for that reason are considered the oldest existing bridges in the world. Given their age, they are in a superb technical condition. They are not large-scale objects—the width of the light of the flow opening is approx. 1.5 m, its height approx. 2 m, **Figure 2**.

It should be highlighted that Mycenaean bridges were elements of the road system, which today is known as *Mycenaean Highways* [8] —which may be a slightly exaggerated name. The term was first introduced by A. Jansen, the author of one of the chapters of this book.

Mycenaean bridges are very interesting from the point of view of structural mechanics and construction technology. At the first glance, they seem to be arch structures. But if so, one must admit that these arches are rather accidental. The arrangements of boulders of which the arches are made demonstrate the lack of knowledge with regard to the essence of the arch behaviour. As a matter of fact, they are corbel structures which—as a result of seismic earthquakes—have been degraded to the present shape. Hence, the currently observed cyclopean boulder arches are a result of the transformation of the corbel system into a possible mixed arch-corbel static system. The effort and stability of Mycenaean bridges were analysed in the paper [9]. Identification of the technology of Mycenaean bridges is a great challenge, **Figure 2**. The cantilever slab technology enables construction without scaffolding. If indeed it was so, Mycenaean bridges were a major constructional achievement. Despite the fact that Mycenaean bridges constitute a part of the heritage of mankind, they have not been sufficiently studied. Paradoxically, apart from taking external measures, their ultra-historical character makes it impossible to conduct standard research.

3. Times of the Roman Empire

The period of the Roman Empire saw an overwhelming abundance of stone, brick and stone-brick arch bridges—apart from military bridge structures, naturally. This period can also be equated with Roman roads. The system and quality of roads enabled movements of the Roman legions and also served civil purposes.

Before we start discussing Roman bridges, however, we need to look at a much older and historically very important road, namely, the Persian Royal Road [10]. It crossed the Persian Empire from Susa or from Persepolis to Smyrna (present-day Izmir). Depending on historical sources, the total length of the road amounted to 2600–3000 km. The road was built by the Persian king of kings, Darius I (c. 550–486 B.C.) and was used as a postal route. It took from 7 to 10 days for the system of courier stations and teams to cover the distance. The road was used by Alexander the Great, and then by the Romans. It was in service for more than 3000 years and coincided with the Silk Road to a considerable degree.

In the town of Diyarbakir (Kurdish: Amed) in south-eastern Anatolia, a bridge called Ten Eye Bridge has been in use till this day, **Figure 3**. It was constructed in the eleventh century, although its dating is connected with the conducted reconstruction. Some historical sources connect this place with the Royal Road and if this were to be the case, it would be one of the oldest bridge

http://dx.doi.org/10.5772/63461

crossings in the world. Assuming that 1065 A.D. is a trustworthy date of the reconstruction, the bridge on the river Tiger (Dicle – tr.) comes from the Roman period.

Figure 3. The 10-Eye Bridge over the Dicle River in Diyarbakir.

Here, the dating of the Roman Empire should be highlighted. In most part of Europe, it covers the period until the fall of the Empire in 476 A.D. In the Mediterranean, it is seen as connected with the fall of Constantinople on 29 May 1453. There is a difference of almost one millennium. In these circumstances, Roman bridges in Turkey and Arab countries may be much younger than Roman bridges in Spain.

One of the first bridges on the river Tiber built in 62 B.C. is certainly Roman—according to the both modes of historical dating, **Figure 4**. There is a funny story connected with the bridge: when it was constructed, the payment for work was refused unless it was proved that the structure was durable. The photograph in **Figure 4** was taken in 2008. Nevertheless, the ordering party's anxiety is easier to understand if we take a look at other, even later, Roman bridges, as shown in **Figure 5**.

a) b)

Figure 4. The Pons Fabricius built by Lucius Fabricius, 62 m long, 5.5 m wide.

The bridge is located on a Roman road bearing a Spanish name: *Via de La Plata*. It is over 700 m long. Its elements include processed granite arches. There are 60 arches based on massive pillars. They are heavy, reliable structures.

The Pons Fabricius marked the beginning of a new way of looking at bridges that, aesthetically, are relatively light and this trend has survived in Rome till this day. With the exception of the Ponte Pietro Nenni that carries an underground line, all the bridges in Rome have arches. Among them, one finds the first reinforced concrete bridge in the world, the Risorgimento of 100 m span, constructed in 1912 by F. Hennebique. Also, in 2011, a beautiful pedestrian steel footbridge with shallow arches was built—the Ponte dela Musica.

Figure 5. Puente Romano de Mérida, built in first century A.D., Guadiana River.

In the Roman period, bridges known as aqueducts were also built, slender and beautiful in their monumentality. The only load they carried was the dead load with an insignificant addition of flowing water, **Figure 6**.

Figure 6. Roman aqueducts, first century A.D.: (a) Los Milagros (the miracles); (b) Segovia aqueduct.

It should be mentioned that during the Roman Empire, but 200 years later, tower aqueducts were constructed. Water flew in a leak-proof pipe located on relatively short arcades. The level differences were solved on the basis of the principle of communicating vessels according to, surprisingly, Pascal's theorem, see **Figure 7**.

http://dx.doi.org/10.5772/63461

Figure 7. Views of the Aspendos aqueduct: (a) from the top; (b) from the ground level.

4. Bridges in the middle ages

During the European Middle Ages, stone arch bridges were continued to be built, although their variety was limited. Also, the road development slowed down. Bridges usually had a defensive character. **Figure 8** shows two bridges leading to the medieval capital of Spain, Toledo, which are a good example of the role and technique of the medieval bridge construction.

Figure 8. The Toledo bridges over the Tajo River: (a) St Martin's Bridge, fourteenth century; (b) Gothic Alcántara Bridge, thirteenth century.

The Toledo Alcántara Bridge should not be mistaken for another Roman bridge of the same name located in the vicinity of the town of Cáceres and constructed in 105 A.D.

In this period in Turkey, a bridge construction canon was developed. Nowadays, such structures are sometimes called Turkish. The bridges in question were still arch bridges, usually made of light colour stones. They have a main span and, possibly, additional spans that serve as flood relief channels in case of high water. It should be born in mind that in the case of the arch bridge, only the arch is the carrying element. The extended walls in arch bridges act only as a façade, very often hiding empty spaces. These spaces were frequently used as utility rooms by bridge guards and sometimes even as guest rooms for travellers. This was the case of the bridge on the river Batman in Turkey, twelfth century, **Figure 9**.

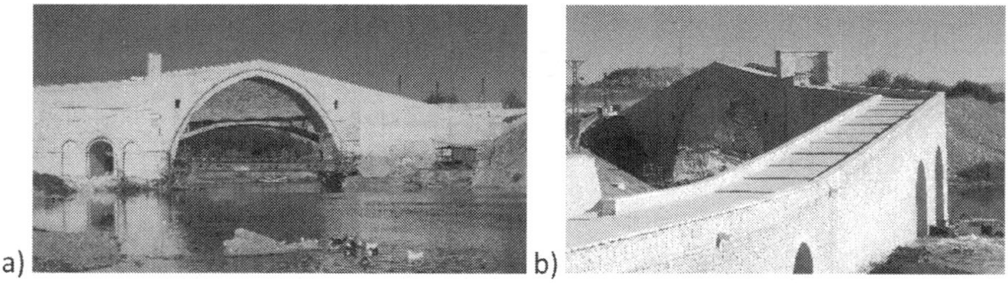

Figure 9. Malabadi Bridge near Sivan: (a) side view; (b) entrance to the bridge room.

A twin object, built by Mimar Hayruddin in sixteenth century, can be found in Mostara on the river Neretva. The stem of the name of the town, which is an adjective, comes from the word *most*, i.e., *bridge*.

Figure 10. Stone arch bridges in the vicinity of Sille Village near Konya: (a) double parabolic arch structure; (b) parabolic arch with the deck partially ruined.

The popularity of arch bridges results from the arch mechanics. To offer an insight into the issue, we can refer to an exercise solved by students in the structural mechanics class. It is about a parabolic three-hinged arch, evenly loaded. Performing basic transform calculations we find that the bending moment at any point of the arch equals nought. It means that, in fact, the arch is not three-hinged but it is a sequence of hinges. Similarly, in the case of a shearing force we find that it equals naught at all the points. It is not a classical approach, but consequently reasoning we can conclude that the arch is a sequence of hinges functioning of which is based on a shear force. Another conclusion is that the only non-zero internal force in the arch is the force normal with regard to its cross-section, and, what is more, it is a compressive force. In

http://dx.doi.org/10.5772/63461

practice, we can shape the sides of stone blocks in such a way that, geometrically, they form a parabolic arch. This arch is going to be a durable and efficient structure—on one condition. In arches, the outward-directed horizontal reaction, called *thrust*, is of primary importance. It has a significant impact and, more often than not, the lack of the proper ground resistance results in the destruction of the arch as a whole or its substantial weakening as a superstructure—at best. The discussed case of a parabolic arch equally loaded is a theoretical one, however, in the case of a real structure we can imagine a set of material points of the highest bearing capacity and that will be an illustration of such an arch. **Figure 10** shows two viaducts on a mountain path near Konya, made with processed stone blocks assembled without mortar.

The advantages of arches in bridge structures make this solution commonly used until this day. The photograph below shows an arch bridge destroyed to such a degree that the arch structure is plainly visible, **Figure 11**.

Figure 11. Structure of stone bridge, Samaria Gorge, Crete: (a) side view; (b) longitudinal view.

Numerous antique bridges were built without mortar. Among the objects shown here there are the Roman Bridge in Mérida and the aqueduct in Segovia. Nevertheless, the columns of the aqueduct of *Los Milagros* are composite, according to modern standards. The external cladding was used as permanent formwork, filled with pozzolana cement concrete.

5. Renaissance: Leonardo bridge over Golden Horn

Renaissance is strongly associated with an outstanding personality of this period, Leonardo Da Vinci. This genius had a brief albeit challenging bridge episode, **Figure 12**.

Through diplomatic channels he was asked to design a project of and construct a bridge over the Golden Horn (Haliç – tr.), an inlet of the Bosphorus in Istanbul. In 1502, an ambassador of Sultan Bayezid II came to Rome. It is probably then, from 1502 to 1503, that Leonardo made the sketch of the bridge that now can be found in the manuscript commonly known as *Paris Manuscript L* [11]. The manuscript pages are rather small, 7 × 10 cm, and the drawing itself is quite simple. Despite the fact that the drawing does not contain any dimensions, it is assumed that the bridge was supposed to be 240 m long and 23 m wide with the arch curvature of 40

m. The bridge was not built, however. In 2001 in Norway, creator Vebjorn Sand constructed a footbridge in the town of Ås [12], which was a reference to Leonardo da Vinci's concept. However—in the author's subjective view—it is only partially successful, both in the aesthetic sense and as an incorporation of the Renaissance genius' idea. Looking at the drawing in the manuscript, it is difficult to decide about the bridge structure. Sand's footbridge seems to be one of many possible interpretations.

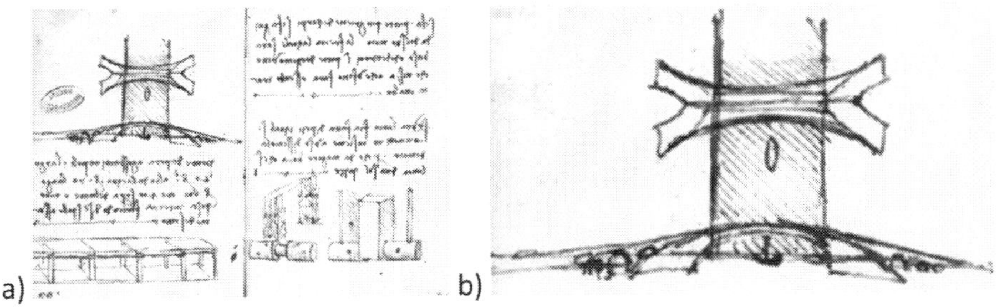

Figure 12. Golden Horn Bridge project, 1502. Paris Manuscript *L*: (a) 66th page of the manuscript; (b) bridge enlarged.

6. Wooden simple bridge

Wooden bridges have always been built. For this reason, a technology was developed which took into account the characteristics of wood as a material of relatively low strength, but at the same time commonly available and easy to handle. The strength of wooden bridges has always been low. It was good if the serviceability period amounted to 20 years. A wooden bridge is also the result of a skilful application of structural mechanics. For the above reasons, usually simple, mechanically pure solutions were chosen. Let us consider the most common wooden bridge, **Figure 13**. Abutments are meant to transfer two kinds of impacts. The first one refers to vertical reactions caused by loads on the carrying deck. For this reason, a row of poles was constructed under each girder. The other impact refers to active ground pressure behind an abutment. Here, the second row of poles was used—outer poles were put behind so-called planking. As a result, a retaining wall was obtained which was also additionally supported by its anchoring in the ground (deadman). The bridge wings were slanting, of variable heights.

An interesting solution is a saddle which enables reduction of support spans and the values of bending moments under the supports. In the case illustrated in **Figure 13**, the bridge beam is single but wooden bolster composite bridge beams were also used. The composition was not complete—from the modern perspective it should rather be called *partial integration*. The construction of wooden bridges was at least as complex as the modern bridge technology. The dimensions of a saddle are determined on the basis of a simple differential calculus assuming extreme curvatures of the beam and the saddle at the local point of contact. Paradoxically, the best work about wooden bridges is a book [13] written 100 years ago. The technology of wooden bridges is still in use in the case of temporary bridges. Additionally, there is a new option available which is connected with the ecofriendly recycled plastic technology, see [14]. Instead of natural or sawn wood, recycled plastic logs or panels are used which contributes to the minimisation of deforestation.

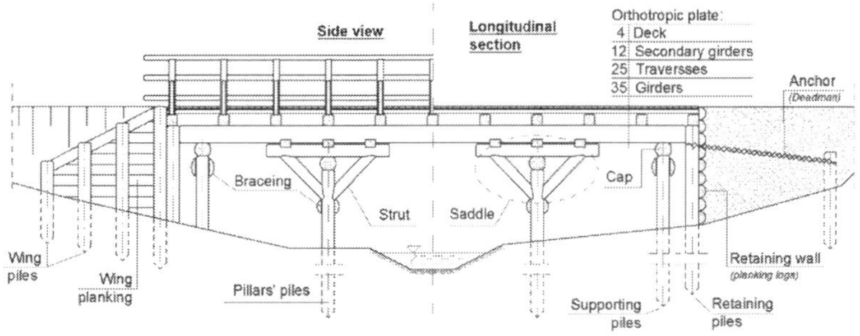

Figure 13. The scheme of a wooden bridge: carrying elements.

7. Iron bridges

The industrial revolution in England was a turning point in bridge construction. In 1779 an arch bridge, known as the Coalbrookdale Bridge or Iron Bridge [15], was constructed by Abraham Darby III with cast iron from his ironworks located in Shrewsbury, **Figure 14a**.

One hundred years later, bridges of spans reaching 160 m were built with wrought iron. **Figure 14b** shows a photograph of the bridge on the river Duro in Porto. It was constructed by T. Seyrig, Gustave Eiffel's associate. At present, the bridge has no utilitarian function, but it is one of the highlights of the city of Porto.

Figure 14. Cast iron bridges: (a) Iron Bridge, 1779 (*photo by Jason Smith* [16]); (b) Dona Maria-Pia Bridge, 1877 (*photo by Dr.Eng. W. Nurek*).

8. Reinforced concrete

Another turning point in bridge construction was the invention of Portland cement and, as a result, using concrete based on it as well as reinforced concrete. The first concrete bridge was a small park object in the botanic garden in Grenoble, **Figure 15**, constructed by Louis Vicat and his son. Vicat was also one on the inventors of cement.

a) b)

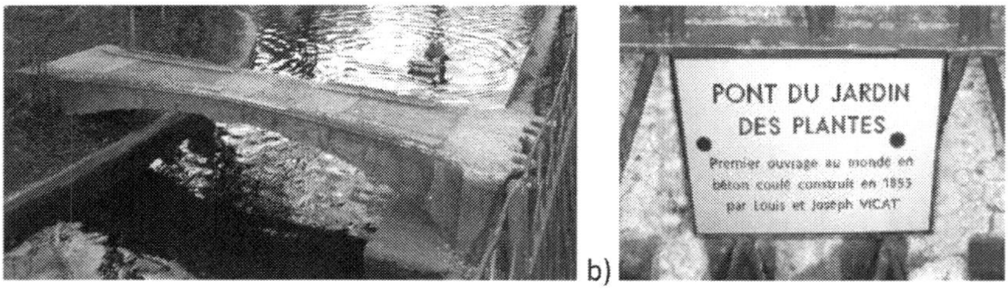

Figure 15. The first concrete bridge, Grenoble 1855: (a) view; (b) memory board (*photos by the courtesy of Prof. Françoise Videau*).

The undisputed *father* of reinforced concrete is Joseph Monier (1867), often mentioned simultaneously with Joseph Louis Lambot who constructed a reinforced concrete boat (1848). Monier gave its surname to the other name of reinforced concrete structures: people spoke of Monier arches or Monier ceilings. In Germany, the name *Monierbau* was initially used. Monier sold his patent to two great engineers, François Hennebique and Gustav Wayss. G. Wayss—the company Wayss u. Freytag—constructed majority of their bridges (about 350 structures) as Monier arches (arched shells) which from the structural point of view copied the stone and brick bridge technology with the only difference that a new material was used, namely, artificial stone: concrete reinforced with bars, originally used for the first time by Monier in the construction of the bridge in Chazelet (1875), see [16].

a) b) c)

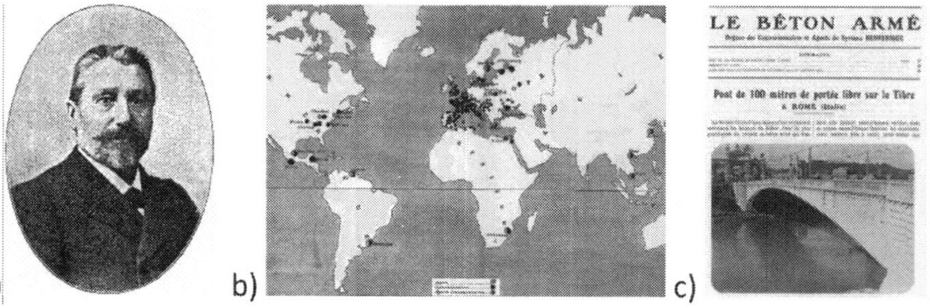

Figure 16. (a) François Hennebique (1842–1921); (b) global activity of *LBA Hennebique*, 1908; (c) the *LBA* issue of February–March, 1912, no. 165–166.

Chronologically, between Vicat's bridge and Lambot's boat there is only an interval of 3 years, but the bridge in Chazelet was built 20 years later. It can be assumed then that the popularisation and implementation of the new technology lasted a similar amount of time.

Even a reinforced concrete bridge had to be invented by someone. Again, there are numerous candidates to the title of the inventor: T. Hyatt (1877), F. Coignet (1861), P. Christophe (1902). It seems, however, that François Hennebique can be indicated as one, **Figure 16a**. He was an author of several patents and, apart from conclusions from the author's bibliographic research, he is named as the reinforced concrete pioneer in various studies, e.g., [17].

Hennebique created a global company (**Figure 16b**) which employed the best engineers and architects building reinforced concrete structures including bridges, **Figure 16**. He offered ready-to-use projects, technology and very often materials as well. From the very beginning Hennebique developed a network of contractors. In practice, the company was known as *Le Système Hennebique* or *Béton Armé Hennebique*. The sale of structural and technological solutions was connected with intensive training and supervision provided by Hennebique.

An essential tool of Hennebique's success in building his company was the technical magazine *Le Béton Armé* [18], published from 1898 to 1939, **Figure 16c**. Apart from purely technical texts in the area of the theory of reinforced concrete as well as the practice, i.e., descriptions of the construction of reinforced concrete structures, one could find information about the network of the company's representatives and contractors. The magazine was richly illustrated with technical drawings, photographs of various construction stages and finished structures as well as advertisements of companies, products and technologies.

Thanks to a happy coincidence, in the city of Lublin in Poland there are two Hennebique beam bridges constructed by Polish engineer Marian Lutoslawski in 1908 and 1909, respectively. One of them has been renovated. After it was put into service, the city's cultural circles took it over and it has been called the *Bride of Culture* since, **Figure 17**.

a)  b) 

Figure 17. Bridge of Culture, Lublin: (a) before restoration, 2011; (b) after restoration, 2013.

The bridges in Lublin are examples of a successful transformation of the bridge technology using wood into the one of reinforced concrete, a new material at the beginning of the twentieth century, see [19].

9. Suspension bridges

The carrying elements of a suspension bridge is a set of pylons and a rope in its natural catenary shape. It is assumed that the origins of suspension bridges are hidden in remote history. The tradition of the construction of narrow footpaths in Asia, especially in Tibet, is widespread. The first modern chain suspension bridge was built over the Menai Strait in 1826, connecting the Welsh island of Anglesey to the mainland. Its maker was brilliant bridge constructor Thomas Telford (1757–1834), see [20].

A turning point in the suspension bridge construction took place at the turn of the nineteenth and twentieth centuries in New York. An engineer constructing suspension bridges in the United States was German immigrant John Augustus Roebling (1806–1869) who also produced wire ropes. Roebling started developing the production of a seven-strand wire rope on a ropewalk that he built on his farm in Saxonburg, Pennsylvania. Next, he built a large industrial complex for the production of wire suited to the needs of the bridges he constructed. His first and significant bridge—Roebling cooperated with two other engineers—was the rail Niagara Falls Bridge which remained in use from 1855 to 1897. The span was 251 m. In 1866, Roebling built a 322-m span bridge in Cincinnati, **Figure 18.**

Figure 18. Roebling's suspension bridge in Cincinnati (1866) [21].

Figure 19. Suspension bridges in New York: left—Brooklin Bridge (1883), right—Williamsburg Bridge (1903) (*photo taken by Alicja Filipowska, 2015*).

Actually, it was a mixed system which nowadays is called the *hybrid suspension and cable stayed bridge*. In a manner of speaking Roebling *repeated* the above- mentioned bridge in New York in

1883. The bridge is called the Brooklyn Bridge. In this case, however, the main span amounted to 486 m. New Yorker Henry Hornbostel (1867–1961) built the Williamsburg Bridge in 1903, see **Figure 19**. This is a road bridge of eight lanes. The longest middle span amounts to 490 m.

Contrary to Roebling, he used a steel frame to raise towers. In this case, the suspension ropes were vertical that became the classical solution for many years. In 1927, Swiss-American engineers Othmar Ammann (1879–1965) and Cass Gilbert (1859–1934) built the double-deck George Washington Bridge. Its middle span amounts to 1067 m. The Mid-Hudson Bridge was built in 1930. The chief engineer was Polish immigrant Ralph Modjeski (1861–1940). The longest span is 910 m, see [22].

The most famous American suspension bridge is naturally the Golden Gate Bridge, constructed in 1937. Its total length amounts to 2737 m, while the main span 1280 m, **Figure 20**.

Figure 20. Golden Gate Bridge, 1937: (a) side view; (b) road traffic (*photo by Alicja Filipowska, 2015*).

At present, the Akashi Kaikyō Bridge (Japan) has the longest central span of all suspension bridges—it is 1991 m long.

All the above-mentioned bridges have truss platforms. It is a proven and reliable solution. Simultaneously, suspension bridges with plate girder deck were constructed such as the Tacoma Narrows Bridge, for instance. The length of its main span amounts to 853 m. The bridge was opened for traffic on 1 July 1940. On 7 November 1940, the catastrophe stuck—and was filmed. The cause of it was the so-called aeroelastic flutter effect. For this reason, the bridge is unofficially known as *Galloping Gertie*. The studies and analyses conducted at that time showed that truss load-bearing structures should be used [23]. Nowadays, when the advanced finite element method (FEM) procedures are available, it is easy to demonstrate by means of numerical analyses that the causes of the catastrophe were correctly identified.

In 1966, the bridge over the Severn Bay near Beachley in Scotland was built. It is a low profile box carrying deck structure. The middle span is 988 m long. An unusual feature of the suspension cables carrying the deck is that they are not vertical, but arranged in a zigzag pattern. The diagonal arrangement of hangers—as compared to vertical ones—increase the dynamic stiffness that is additionally supported by tuned mass damper-harmonic absorbers.

The deck is an orthotropic steel box of the aerofoil shape. The bridge was tested and approved by the designers Freeman, Fox and Partners following wind tunnel tests.

The First Bosphorus Bridge constructed in Istanbul in 1973 is a replica of this bridge. The main span—the distance between the towers—is 1074 m, **Figure 21**. When opened, it was the second bridge crossing between Europe and Asia constructed since 480 B.C.

a) b)

Figure 21. First Bosphorus Bridge: (a) side view; (b) tower and aerodynamic low profile deck.

Currently, the construction works of the third Bosphorus bridge—Yavuz Sultan Selim Köprüsü —are underway, according to the design by Michel Virlogeux and Jean-François Klein. The bridge has been designed as a hybrid structure.

10. Cable-stayed bridges

A cable-stayed bridge has one or more towers, from which straight cables carry the bridge deck as elastic supports. There is a clear analogy between the behaviour of backstays and cable-stayed bridges. The first known analysis of this problem comes from 1823 and was performed by C.L. Navier, see [24]. Diagonal rods in Bollman trusses also bring to one's mind cable-stayed

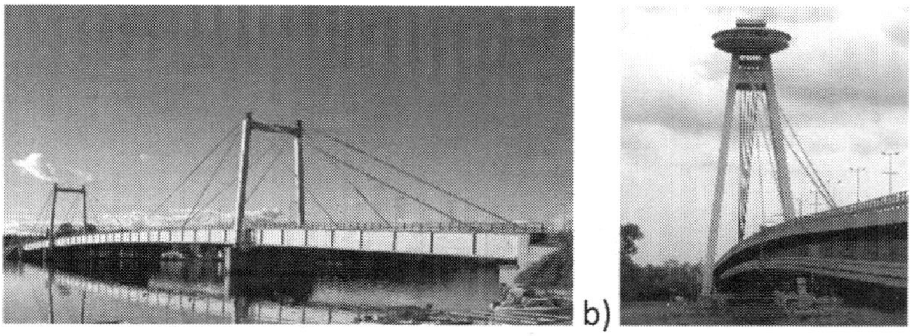

a) b)

Figure 22. Cable stayed bridges: (a) Strömsund Bridge (1956) (*photo by Lars Falkdalen Lindahl* [25]); (b) the tower of the Bratislava SNU bridge, 1972.

bridges. In 1873 in London, the Albert Bridge, designed by Ordish and Bazalgette, was constructed. The first modern cable-stayed bridge is the Strömsund Bridge in Sweden, designed by Franz Dischinger (1956), **Figure 22a**.

Considering the development of the cable-stayed structure, one can specify the German period connected with German engineers F. Dischinger, U. Finsterwalder , F. Leonhardt, H. Homberg, H. Wittfoht, E. Jux and others. Naturally, cable-stayed bridges have been constructed everywhere. **Figure 22b** shows one on the most beautiful bridges of this kind, namely, the Bridge of the Slovak National Uprising in Bratislava, constructed in 1972. Another spectacular cable-stayed bridge is the bridge located at the outlet of Lake Maracaibo, designed by R. Morandi and built in 1962, see [26].

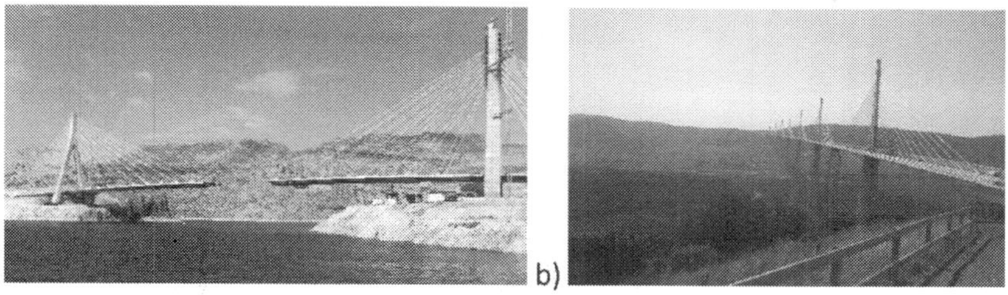

Figure 23. Stayed cable bridges: (a) segmental assembling of the Nissibi Euphrates Bridge, 2015; (b) Norman Foster's Millau Bridge (*photo by A. Leniak-Tomczyk, 2004*).

In the case of cable-stayed bridges at least two assembly technologies are available. The first one refers to a segmental bridge structure built in short sections. It is a very natural method of constructing such bridges. Successive segments are stabilised by attaching them to pylons by means of cables, **Figure 23a**. The other method consists in the incremental launching of a carrying-deck with the use of fixed and intermediate supports. After the launching, the bridge is rectified to its proper grade. An example here is the architecturally magnificent Millau Viaduct designed by M. Virlogeux and N. Foster (2004), **Figure 23b**.

Figure 24. Visible cables: (a) the Golden Horn Metro Bridge, 2014; (b) extradosed Viaducto de Teror, Gran Canaria, 2010.

Cable-stayed bridges are aesthetically challenging. To highlight their attractiveness appropriate lighting is required, **Figure 24a**. **Figure 24b** shows a bridge in the mountainous part of Gran Canaria where an austere structure matches an equally austere mountainous landscape. The whiteness of the bridge contrasts with the surroundings. The backstays are clearly visible—paradoxically, thanks to cloudy weather. Very often backstays are not discernible at all and for this reason, at night special illumination is used.

In the family of cable-stayed bridges, a special group can be distinguished on constructional grounds, namely, *extradosed* bridges, **Figure 25b**. In this case, the inclination of cables measured from the deck level to the cables is significantly lower than $\Pi/4$. Projecting the normal force acting in the cable N onto horizontal and vertical directions we arrive at N_H, N_V components and, additionally, we obtain $N_H > N_V$. As a result there occurs a significant compression in the carrying-deck which in the extreme degree is manifested near the tower. The horizontal force can be used in the design as the force pre-stressing the deck longitudinally.

Figure 25. Badajoz, Spain: (a and b) views from the Roman Bridge.

During the design works and the construction of the Golden Horn Metro Bridge in Istanbul, there was an on-going global discussion about limiting the architectural dominance of a constructed bridge, see [27]. Despite the fact that the pylons demonstrated Ottoman features it was suggested that, with regard to the global cultural and architectural heritage of the area in the vicinity of Golden Horn, the bridge would constitute an extraneous dominant. The bridge was completed without any changes in 2015, see **Figure 24a**.

Short distances between bridges in cities contribute to amusing and sometimes even grotesque situations. Two gorgeous bridges in Sevilla can serve as an example. The Puente de la Barqueta is a tied-arch bridge designed by J. Arenas and J. Pantalerón. The other bridge is the cable-stayed Puente del Alamillo designed by S. Calatrava. The construction of the both bridges was completed in 1992. They are located on a straight strip of the oxbow lake of the river Guadalquivir, 1 km apart. Their views overlap, which can be rather irritating.

It is quite a common case in highly urbanised areas and, actually, it occurs in every city on a big river. **Figure 25** shows two frames of a movie that is, so to speak, created in the head of a passer-by walking on the Roman Bridge over the Guadiana River in Badajoz. Depending on the spectator's mood and perception it can appear as chaos or an interesting coincidence.

Cable-stayed bridges are efficient in terms of bridge structure mechanics when placed between cantilever and suspension bridges. On this position, they also prove to be economic solutions. It means that they can be used in the area of small architecture, even if beam or plate bridges are cheaper and better, in a sense. Above all, the pylon, as an interesting dominant, contributes significantly to the attractiveness of a local landscape, see **Figure 26**.

Figure 26. Architectural cable-stayed bridges: (a) the tram bridge in Bydgoszcz (*photo by courtesy of Gotowski Company*); (b) the bridge in Dźwirzyno over the Resko Channel (*photo by M. Delmaczyński*).

Figure 26a shows a tram bridge on the river Brda in Bydgoszcz, 75 m long. It was designed by K. Maciejewski and constructed in 2014. Among typical, ordinary urban buildings the short pylon is a moderate dominant contributing nevertheless to making the landscape more interesting when contrasted with the dynamics of the inclined pylon. Similar enhancement is visible in **Figure 26b**. The short pylon of a small, 51 m long bridge, located in the village of Dźwirzyno, is so different from the village buildings that it becomes a fascinating radical sculpture. It was designed by J. Siuda and M. Delmaczyński. After its construction in 2011, an increased demand for equally small but architecturally interesting bridges has been observed.

Figure 27. Arch bridge erection by cable-stayed supports at sunset, Estremadura, 2015.

The cable-stayed technology has been used as a temporary support for arch bridge scaffolding for many years. These are transitional arrangements and images but ones that are truly charming thanks to the additional spaciousness they create. Sometimes, as shown in **Figure 27**, an austere surrealist image is obtained. This short-lived aesthetic form refers directly to the art of performance as it disappears the moment the arch is built.

Figure 28. Katehaki footbridge, Athens, 2004: (a–c) different views of the bridge.

A separate place in architecture and structural engineering belongs to Santiago Calatrava, although in this paper he is mentioned only briefly. Calatrava, in the architectural millieu's opinion, is a creator of gigantic forms—spatial sculptures which enhance the landscapes of various cities around the world. As a bridge constructor, he broke a mental barrier existing in the area of bridge design where bridges were designed for bending as a dominant mechanical state. In the case of pedestrian footbridges he constructed bridges where carrying elements are screwed together. Due to this, the mental barrier has been broken.

Examples of gigantic sculptures in urban areas include the following bridges: the Puente del Alamillo in Sevilla, 1992, the Puente De La Mujer in Buenos Aires, 2001 and the Sundial near Redding, California, 2004. The Jerusalem Chords Bridge, however similar, is new concept. These bridges are purely white, the pylons are inclined by approx. 50° and they resemble each other, as images, to a high degree. However, the structural engineering of each of the bridges is different. The Alamino is a road bridge with a beam load-bearing structure identical to the one of the arch bridge Puente Lusitania in Mérida. The rest are pedestrian footbridges with screwed load-bearing structures. The Puente de la Mujer is a moveable bridge with a rotary movement about the vertical axis on the support with a pylon. The Sundial has a truss load-carrying structure. As demonstrated, each of these bridges is different and only for architects their images are identical.

In the author's opinion, the most interesting is the Katehaki Footbridge in Athens, 2004, **Figure 28**. The footbridge is never entirely visible and therefore mysterious—making one yearn for more. The pylon is slightly bent as a result of which it loses the original but at the same time primitive form of an opened set square and becomes similar to boats crossing the Mediterranean Sea. The footbridge is literally squeezed between uninteresting street buildings of Athens. A crossing passer-by who reaches one of its ends can look into the windows of the buildings located only 10 m away. The footbridge is in the middle of street traffic. The lack of space and the constant flow of passing vehicles contribute to the dynamics of the structure which, thanks to its white and slender elegance, floats above crowded streets of Athens.

The Millennium Bridge in London is a hybrid of two static schemes. There are deck segments supported by cables (*vide* the ribbon scheme) and at the same time the deck is suspended on the same cables as in the case of the suspension bridge. **Figure 29**.

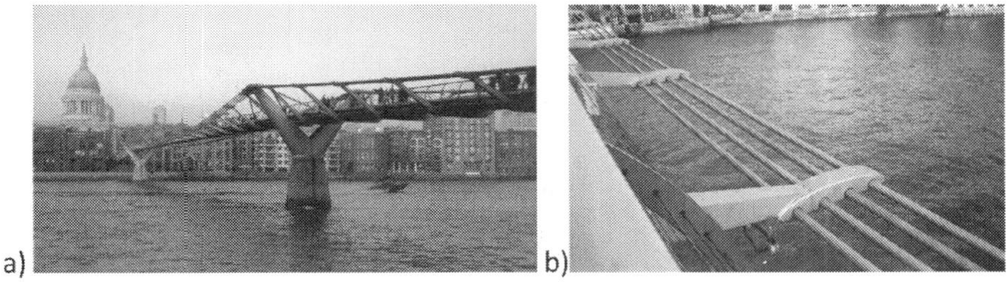

Figure 29. Millennium Bridge in London: (a) view of the bridge; (b) support detail.

The history of the bridge opening is a typical English story, i.e., starting from a total failure and ending in full glory, see, e.g., [28]. The bridge was conceived as an interesting design, quite innovative.

The opening day had been expected for a long time. Finally, on 10 June 2000, thousands (ca. 5000) of people were waiting to enter the bridge and walk to other side. At this moment, the new bridge entered into a state of unpleasant complex horizontal and vertical vibrations. It turned out that despite the use of advanced computational techniques and the designers' experience, the bridge demonstrated dynamical over-sensitiveness. The new bridge was closed on 12 June of the second millennium. Soon after, a diagnosing research was conducted which pinpointed the cause of the dynamic instability of the bridge. On its basis, it was decided to install a system of dampers which changed the dynamic response of the bridge. Two years later, the Millennium Footbridge was reopened and has been working properly ever since.

11. Instead of a summary

There are many bridges, constructors and technologies that should be and are described and discussed in various papers and monographs. From the wide range of existing bridge structures, one conclusion can certainly be drawn: every bridge is important.

Author details

Slawomir Karas

Address all correspondence to: s.karas@pollub.pl

Lublin University of Technology, Faculty of Civil Engineering and Architecture, Road and Bridge Department, Lublin, Poland

References

[1] Sustainable bridges. (2007). In J. Bien, L. Elfgren, J. Olofson (Eds.), Assessment for Future Traffic Demands and Longer Lives. Dolnoslaskie Wydawnictwo Edukacyjne, ISBN 978-83-7125-161-0.

[2] Vallero D., Brasier C. (2008). Sustainable Design: The Science of Sustainability and Green Engineering. John Wiley and Sons, Inc., Hoboken, NJ, ISBN 0470130628. http://books.google.com/books/about/Sustainable_Design.html?id=vrj5DIQ-7m8C%7C [19.01.2016].

[3] Lesaffer R.C.H. (2011). Kellogg-Briand Pact (1928). In R. Wolfrum (Ed.), Max Planck Encyclopedie of Public International Law.Oxford University Press, Oxford. https://pure.uvt.nl/portal/files/1336402/Lesaffer_KelloggBriandPact_110617_post-print_immediately.pdf [22.01.2016].

[4] Eurocode—Basis of structural design.

[5] Karas S., Hypki M. The changes of bridge loads during the long-term period. KSCE Journal of Civil Engineering. DOI 10.1007/s12205-015-0355-7 [17.02.2016].

[6] The history of Herodotus translated into English by G.C. Macaulay. http://wps.pearsoncustom.com/wps/media/objects/2426/2484749/chap_assets/bookshelf/herodotus.pdf [02.02.2016].

[7] Nebel B. Julius Cäsars Brücke über den Rhein. http://www.bernd-nebel.de/bruecken/index.html?/bruecken/3 bedeutend/caesar/caesar.html [07.02.2016].

[8] Echos du Monde Classique. Classical Views. Classical Association of Canada, University of Calgary Press, Vol. 41, p. 3, 1997. http://archhades.blogspot.com/2012/11/ancient-greek-art-mycenaeans.html [04.02.2016].

[9] Karas S., Kowal M. The Mycenaean bridges—technical evaluation attempt. Roads and Bridges—Drogi i Mosty. Vol. 14, No. 4, 2015, pp. 285–302. DOI: 10.7409/rabdim.015.019 [11.01.2016].

[10] Wiesehöfer J. (1996). Ancient Persia, I.B. Tauris Publishers. https://ia801406.us.archive.org/14/items/AncientPersiaFrom550BcTo650Ad2001/JosefWiesehoferAncientPersiaFrom550BcTo650Ad2001.pdf [13.01.2016].

[11] File:Leonardo Bridge.jpg [11.02.2016].

[12] Leonardo's Bridge. http://structurae.net/structures/data/index.cfm?ID=s0002393 [11.02.2016].

[13] Melan J. (1922). Der Brückenbau I. Bild.: Einleitung und hölzerne Brücken. 3. Auflage Leipzig u.Franz Deuticke, Wien.

[14] Chandra V., Kim J.S., Nosker T.J., Nagle G.J. World's First Thermoplastic Bridges. http://theinfrastructureshow.com/audio/downloads/Worlds-First-Thermoplastic-Bridge.pdf [17.02.2016].

[15] File:Ironbridge002.JPG | Ironbridge002 [17.02.2016].

[16] Chazelet Bridge. http://structurae.net/structures/chazelet-bridge [10.02.2016].

[17] McBeth D.G. (1998), Francois Hennebique (1842–1921) – reinforced concrete pioneer, Proceedings of the Institution of Civil Engineers, Vol. 126, pp. 86–95.

[18] Universiteitsbibliotheek Gent. http://lib.ugent.be/lebetonarme/about-journal.html [15.013.2016].

[19] Karas S (2013). Unique Hennebique Bridges in Lublin, Poland. American Journal of Civil Engineering and Architecture, Vol. 1 No. 2, pp. 47–51. DOI: 10.12691/ajcea-1-2-5 [17.02.2016].

[20] Cragg A.P. A Study of the Menai Suspension Bridge. Proceedings of Bridge Engineering 2nd Conference, 2008, Bath, UK. https://www.thomastelford.com/books/SampleChapters/Thomastelford250samplematerial.pdf [30.03.2016].

[21] File:RoeblingBridge1870.jpg | RoeblingBridge1870 [10.02.2016].

[22] Ralph Modjeski. https://structurae.net/persons/ralph-modjeski [10.02.2016].

[23] Historic American Engineering Record Tacoma Narrows Bridge, HAER No. WA-99 http://lcweb2.loc.gov/master/pnp/habshaer/wa/wa0400/wa0453/data/wa0453data.pdf [10.02.2016].

[24] Gimsing N.J. History of cable-stayed bridges. Proceedings IABSE Conference Cable-Stayed Bridges: Past, Present and Future, Malmö, Sweden, 2?4 June 1999. http://

www.ce.jhu.edu/perspectives/protected/handouts/IABSE%20cable%20stayed %20bridge%20article.pdf [12.01.2016].

[25] File:Strömsundsbron 2010.jpg [11.01.2016].

[26] General Rafael Urdaneta Bridge. https://structurae.net/structures/general-rafael-urdaneta-bridge.

[27] Loos M., Korus Ch., Tebard Th., Wachten K. Independent Assessment of the Visual Impact of the Golden Horn Metro Crossing Bridge on the World Heritage property "Historic Areas of Istanbul". http://services.arch.rwth-aachen.de/forschung/viar-istanbul-2011-en.pdf [02.02.2016].

[28] Dallard P., Fitzpatrick A.J., Flint A., Le Bourva S., Low A., Ridsdill Smith R.M., Willford M. The London Millennium Footbridge. The Structural Engineer, Vol. 79, No. 22, 2001, pp. 17–33. https://researchcourse.pbworks.com/f/structural+engineering.pdf [30.03.2016].

Chapter 2

Highway Bridge Traffic Loading

Alessandro Lipari

Additional information is available at the end of the chapter

http://dx.doi.org/10.5772/63967

Abstract

In this chapter, traditional approaches and recent advances in highway bridge traffic loading are described, which are of great significance for structural safety assessment of bridges. Indeed, it is widely accepted that consideration of site-specific traffic features can enable significant savings in maintenance operations. While short spans are governed by free-flowing traffic plus an allowance for the dynamic effects, long spans are governed by congested conditions. For the former, a promising research trend is the investigation of the dynamic vehicle-bridge interaction, which is shown to lead to dynamic effects much lower than previously thought. For the latter, advances in traffic flow modelling enable the simulation of realistic congestion patterns based on widely available free-flowing traffic data, thus partially overcoming a long-standing shortage of congestion data. Here, emphasis is given to the promising application of traffic microsimulation to long-span bridge loading, combined with a probabilistic approach based on the extreme value theory, to compute site-specific characteristic loading values.

Keywords: highway, traffic loading, microsimulation, probabilistic modelling

1. Introduction

Highway bridges have the main objective of carrying traffic; therefore traffic loading is a primary source of stress on the structure – quite often the largest one. However, traffic conditions are rather variable, as they depend on many factors, ranging from the macroscale (such as economic development) to the microscale (such as individual driver behaviour). While the resistance of bridges has been fairly well studied, less attention has been paid to the load to which a bridge is actually exposed.

Not surprisingly, the weight of heavy vehicles is of particular interest in traffic loading. Maximum weights are regulated by national governments and the truck weight allowance has constantly increased over the years. Not only have the single truck weights increased, but also the number of trucks on the road has steadily grown. Furthermore, overloaded and non-regulated trucks are not such rare events. Therefore, while the road infrastructure is inevitably deteriorating, the load to which it is exposed is globally increasing.

To account for the large variability of vehicle weights and traffic conditions, codes of practice prescribe fairly conservative load models for the design of new bridges, whereas only few codes are available for the assessment of existing bridges. Furthermore, the vast majority of codes are limited to short and medium spans.

On the contrary, most bridges are not likely to experience the high level of load prescribed in the design codes, with the consequence that the applied design load models may be disproportionate to the traffic that the bridge actually carries. This approach is generally acceptable for new bridges, for which an increase in load typically requires a less than proportionate increase in construction costs, whereas in the case of existing bridges it may play a decisive role in planning maintenance operations [1]. This may even result in the bridge being replaced unnecessarily or prematurely.

It is therefore apparent that the safety conditions of existing bridges need to be carefully reassessed to avoid unsafe situations or else unnecessary maintenance. Nowadays, it is relatively easy to obtain information on the traffic expected to occur on an existing bridge. The use of such site-specific traffic data may enable tailored maintenance operations, thus leading to an optimal – yet safe – use of the infrastructure. Significant savings can be achieved in both economic and environmental terms (e.g. saved maintenance costs and material production, or avoided congestion due to traffic disruptions).

1.1. Research on highway bridge traffic loading

Research in bridge loading is often related to the development of codes or standards. In the context of bridge loading, it is convenient to define a short-span bridge as a bridge whose governing traffic case is free-flowing traffic plus an allowance for dynamic effects, as opposed to a long-span bridge, which is governed by congested traffic with no dynamic effects. The bridge length threshold between the two cases depends on many factors but it is currently thought to lie between 30 and 50 m.

Research on highway bridge traffic loading has mainly focussed on short-span bridges. For those bridges, The governing traffic case typically consists of or two big vehicles in free-flowing conditions, which dynamically interact with the bridge. No cars are involved in the governing case; hence, data about individual heavy vehicles generally suffice. This information is nowadays commonly available from *Weigh-In-Motion* (WIM) stations, often paired with *inductive loop* detectors. Importantly, recent studies have shown that the dynamic increment for extreme loading events may not be as high as previously thought [2, 3]. This has the potential of lowering the above-mentioned threshold between the two governing traffic states [4].

In contrast, long-span bridge loading is governed by congested traffic. Vehicles strongly interact with each other and driver behaviour becomes relevant. Cars cannot be neglected, as

http://dx.doi.org/10.5772/63967

they play an important indirect role by keeping heavy vehicles apart. Unfortunately, there is a long-standing shortage of congested traffic data, mainly due to current limitations of detection techniques. This is reflected in the fact that most existing long-span bridge traffic load models are based on conservative assumptions, such as a queue of vehicles at minimum bumper-to-bumper distances [5–11], thus neglecting driver behaviour. However, traffic-related technologies are developing rapidly, thus enabling a better understanding of driver behaviour and overall traffic features, particularly during congestion. It is therefore sensible to introduce both recent and consolidated advances in traffic modelling into bridge-loading research. Among those, traffic *microsimulation* is a powerful tool to simulate realistic congested scenarios, based on widely available free-traffic measurements. Furthermore, increased computer performance allows for the simulation of the long periods required to identify extreme loading events.

1.2. Methodology

In general, the process to compute site-specific bridge traffic loading consists of the following steps:

1. Traffic data collection: This provides the basis for the analysis. It traditionally includes truck weight and axle data, and more recently vehicle speed and time headways, generally sufficient for short-span bridges. However, reasonable assumptions need to be made during congestion, due to the shortage of congested traffic data. Section 2 introduces some traffic engineering concepts, with which bridge engineers may not be familiar.
2. Generation of a database: As the traffic Data is often not large enough to identify the rare loading events used for bridge design and assessment, an extended *garage* of fictitious vehicles may be generated using common Monte Carlo techniques based on the recorded data.
3. Simulation of load effects: The traffic database is passed over a bridge and the required load effects are computed, for instance using influence lines or finite element analysis; if relevant, dynamic effects are also computed.
4. Extrapolation: As it may be still quite computationally demanding to simulate traffic for very long periods, load effects are further extrapolated to find *characteristic values* with the safety level required by the codes of practice. These values can be then compared to the resistance of the corresponding members. Section 3 gives an overview of extreme value statistics, as it is an appropriate tool to find characteristic load effects. Sections 4 and 5 present bridge loading respectively for short and long spans, highlighting consolidated results and current research trends; emphasis is given to the combined application of traffic microsimulation and extreme value statistics to find characteristic loading values for long-span bridges.
5. Model calibration: When the target is to develop a code to be applied for a range of conditions, then a *notional* load model is found which envelopes the considered load effects. The calibration process often includes *reliability analysis* to derive appropriate partial safety factors. This step is not dealt with in this chapter; the reader can refer, for instance, to Refs. [12] or [13] for further details.

2. Traffic engineering concepts

The most common traffic characteristics are *flow* (sometimes called *flux* or *volume*), *density* (sometimes called *concentration*) and *mean speed*. Flow is inherently a *temporal* quantity (number of vehicles per unit of time), density a *spatial* one (number of vehicles per unit of length) and the *mean speed* can be either, depending on whether speed is averaged at a certain point over a time interval (*time mean speed*, v) or at an instant of time over a stretch of road (*space mean speed*, v_s). Density is a key traffic variable for bridge-loading applications, as it is directly related to the number of vehicles present on a bridge at any one time. Speed is related to minimum inter-vehicle gaps: the lower the speed, the smaller the minimum safe distance between vehicles.

There are two main detector types for collecting traffic data: *point* detectors, which count the number of vehicles in a unit of time (the natural way to collect *flow* data) and *spatial* detectors, which count the number of vehicles in a unit of length (the natural way to collect *density* data). In practice, flow is far easier to measure than density, as it can be measured by means of common point detectors such as induction loops [14].

Clearly, the knowledge of vehicle positions is a prerequisite for any subsequent structural analysis. However, as traffic Data is available only at selected road cross-sections, the actual vehicle positions along a stretch of road can only be estimated from such point measurements, typically assuming constant speed. As will be shown later, this is a reasonable assumption in free-flowing traffic and therefore appropriate for short-span bridges. However, during congestion (relevant to long-span bridges), speeds may vary significantly, like in the common case of *stop-and-go waves*. In this case, the estimation of vehicle positions from point measurements may result in a significant loss of accuracy [15].

2.1. Traffic theory

The Fundamental Equation of Traffic (FET) has been long used to relate *flow* q, *density* k and *space mean speed* v_s [16, 17]:

$$q = k \cdot v_s \tag{1}$$

Eq. (1) implicitly assumes that each vehicle maintains a constant speed, although individual speeds may be different. Given the large availability of point measurements, density is typically estimated from Eq. (1) from flow and speed data.[1] Even when vehicles do not keep their speed, the FET might still be able to provide fairly accurate density estimates during congestion [18].

---

[1] The space mean speed v_s can be reasonably approximated as the harmonic mean speed of individual vehicles collected at one point, although it is by definition a spatial quantity. The use of the time mean speed (arithmetic mean) in Eq. (1) is incorrect, although frequent, leading to a systematic underestimation of the actual density when traffic is not stationary, that is when there are large variations of speed, typical of most congestion events [18].

http://dx.doi.org/10.5772/63967

Single-vehicle data is usually aggregated over a time interval varying from 20 seconds to 5 minutes. Aggregated *macroscopic* variables, such as those implied in Eq. (1), are useful to obtain a global and concise description of the traffic stream. However, for bridge-loading applications, it is highly desirable to also have single-vehicle *microscopic* data, such as time stamps, so as to identify vehicle configurations and reconstruct vehicle positions.

The motion of individual vehicles can be fully described by tracing their trajectory, plotted over space-time domains of the traffic stream, such as in the example of **Figure 1**.[2] Space-time domains can be generated with a rapid sequence of aerial photographs or a video [21, 22], or else with a dense installation of loop detectors or other point sensors [23]. In fact, both options are rarely practicable and currently limited to research applications.

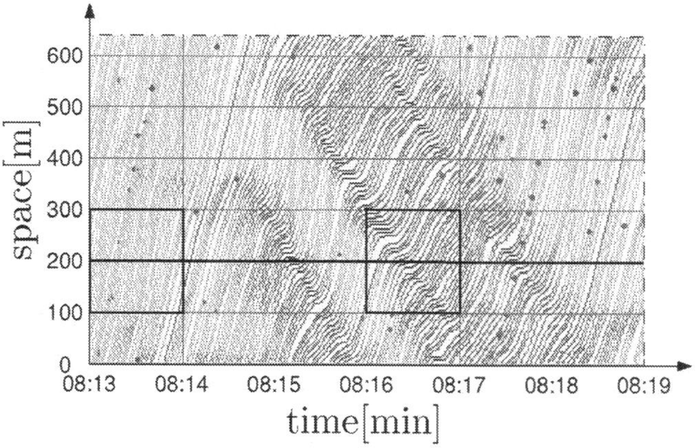

Figure 1. Example of vehicle trajectories collected on the US-101 highway near Los Angeles (adapted with permission from [24]).

Let us consider the common case of a point detector. It would collect traffic data as per the straight line depicted at 200 m in **Figure 1**. Say the detector provides both macroscopic data aggregated every 60 s and microscopic single-vehicle data. Let us assume we are interested in inferring the density over a stretch of 100 m either sides of the point detector. In the first minute (08.13–08.14), the traffic stream flows quite steadily at about 45 km/h. Therefore, Eq. (1) applies and it is then possible to accurately infer density over the 200 m length, as well as individual vehicle positions from single-vehicle data. Afterwards, the traffic flow breaks down, developing *stop-and-go waves* (a stopped vehicle can be recognised when its trajectory is horizontal in the space-time domain). In the time interval 08.16–08.17, it may be possible to compute a reasonably accurate *average* density estimate over the 200 m length from Eq. (1), but this is likely to miss the critical *maximum* density occurring at some point within the 60 s aggregation

[2] Edie [19] generalised the FET for any space-time domain (like the one in **Figure 1**), making it adaptable to any kind of detector and traffic state. Edie's equations are exact and consistent by definition, but only within the regions where measurements can be effectively taken, for instance, density between two closely spaced loop detectors [20].

time. Even when single-vehicle data is available, it is not possible to readily reconstruct the spatial distribution of vehicles, as vehicle speeds vary from those recorded when the vehicles crossed the point detector.

2.2. Data collection

Traffic weight data is traditionally based on roadside truck surveys, or, more recently, WIM measurements. High-speed WIM stations are able to weigh axles and collect time stamps without stopping vehicles. Axle time headways can be then computed. *Double loop detectors* are used for supplementing information regarding speed. This enables the distance between axles to be computed and to reconstruct the vehicle configuration. The overall vehicle length can also be detected.

Data from paired WIM and loop stations has been widely used. Single-vehicle data is normally available for those stations, although sometimes only for heavy vehicles. Unfortunately, many WIM and loop detectors are not currently reliable at very low speeds [14]. As a consequence, data is largely collected during free-flowing traffic conditions, which also occur more frequently than congested conditions, whereas data about slow-moving vehicles is generally lacking.

The recorded traffic data may be directly used for subsequent structural analysis. Nevertheless, since the recorded dataset is rarely sufficiently large, it is preferable to use it as a basis to generate additional artificial traffic by means of common Monte Carlo simulations for further use in structural analysis – see for instance [25].

Except for very short spans, the next step is to reconstruct the spatial distribution of vehicles from the recorded point measurements. This is equivalent to find the *headways* or gaps between vehicles.[3] When using individual recorded speeds, the headway h between the current vehicle i crossing the detector at a time stamp t_i and the leading vehicle i-1 (that crossed the detector at a time stamp t_{i-1} and with speed v_{i-1}) can be estimated as follows:

$$h_i = v_{i-1} \cdot (t_{i-1} - t_i) \tag{2}$$

thereby assuming that the leading vehicle is keeping its speed v_{i-1}.[4] As seen in Section 2.1, this is an acceptable assumption only in free-flowing traffic. However, when load effects are calculated during congestion, large variations in speed may result in unrealistic spatial distributions or even vehicle overlapping. This aspect is particularly significant for long spans [15].

[3] Here the headway is intended as the distance between same points of two consecutive vehicles, for instance, the front axle or the front bumper; the gap is the bumper-to-bumper distance between two vehicles. They can also be intended as time distances, and when so it will be specified. Note that different interpretations of these terms may be found.

[4] Other assumptions on the speed are also possible, and these will clearly affect the estimated headway. For instance, some studies assume a constant speed for all the vehicles in the traffic stream [26, 27], although the recorded speeds were different. This implies that vehicles are passed on the bridge with the same time headways as were recorded, but not with the same space headways, which depend on the assumed speed. For a more detailed discussion, see [15].

In theory, the use of cameras over a stretch of road would provide accurate information about vehicle positions. However, there are several practical issues which make a camera-based approach difficult, such as sensitivity to lighting conditions or heavy post-processing requirements [14]. Cameras have been deployed for research purposes [21, 22, 28–30] and are becoming increasingly popular, thanks to the recent technological advances. In bridge-related studies, only a few studies report that cameras were used to collect traffic data [7, 11, 31–33].

2.3. Congestion

In simple terms, congestion forms whenever the *inflow* Q_{in} (demand) is greater than the *dynamic capacity* Q_{out} (supply). In reality, inflows greater than the dynamic capacity Q_{out} are possible and the maximum flow that can be attained is named *static capacity* Q_{max}, or simply *capacity*. However, in this case, the traffic flow is not stable and a significant perturbation (e.g. a braking vehicle) could break the flow down and generate a queue (*stop-and-go wave*), propagating backwards at a typical speed of 15 km/h. The flow coming out of such a queue is namely the dynamic capacity or *queue discharge rate*. The inflow Q_{in} can be easily collected from point detectors, whereas several procedures are available to estimate the capacity Q_{max} (see for instance [34]). The capacity depends on many factors, such as the road geometry; importantly for bridge-loading applications, it also depends on the truck percentage. The estimation of the dynamic capacity is not as straightforward, but research suggests it is 5–10% less than the static capacity Q_{max} (see for instance [35]).

Both capacities can be further reduced by *bottlenecks* due to a variety of causes, as will be discussed in the next section. The *bottleneck strength*, ΔQ, can be defined as the difference between the dynamic capacity in normal conditions, Q_{out}, and the reduced dynamic capacity when a bottleneck is in place, Q'_{out}:

$$\Delta Q = Q_{out} - Q'_{out} \tag{3}$$

Depending on the inflow Q_{in} and the bottleneck strength ΔQ (and for a given traffic history), the traffic can take up any of the traffic states outlined in **Table 1** [36, 37]. Combinations of congested states may also occur.

In general, increasing inflow and/or bottleneck strength has the effect of moving down the table to a higher intensity of congestion. In addition, the greater the bottleneck strength, the lower the average speed and the lower the speed oscillations during congestion [38]. Congested states that occupy a significantly long stretch of road (so-called *extended* states), such as SGW, OCT and HCT, are of particular significance for long-span bridge-loading applications. For comparison with the common traffic loading assumption, the full-stop condition (FS) is also included.

Spatio-temporal speed plots are useful for visualising congestion patterns (**Figure 2**). The space mean speed is collected at four virtual detectors and aggregated over 60 s. **Figure 2(a)** shows a SGW state, where the waves are clearly visible as peaks. **Figure 2(b)** shows a combined HCT/

OCT state, where the upstream small oscillations typical of the OCT state fade away into a HCT state downstream, where there are essentially no oscillations.

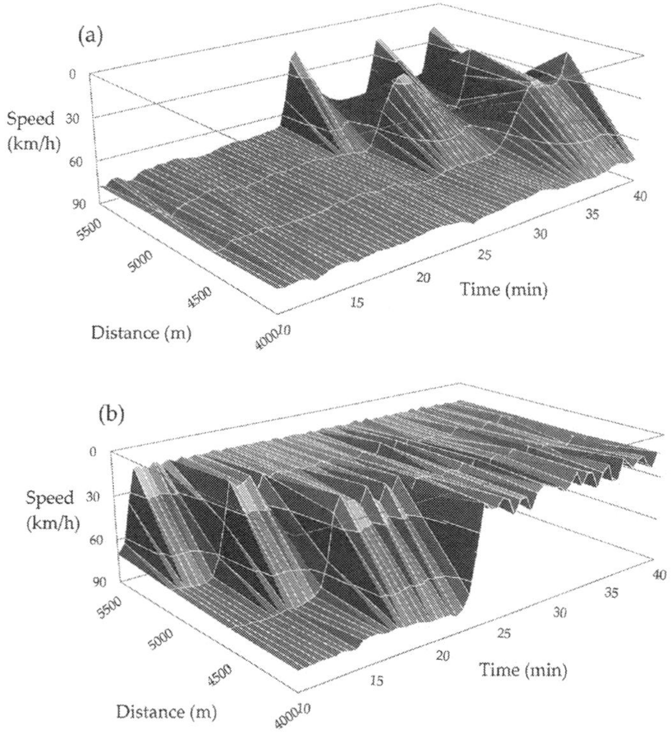

Figure 2. Spatio-temporal speed plots for a simulated two-lane flow with 20% trucks and bottleneck strength: (a) 270 and (b) 1056 veh/h.

2.3.1. Causes and effects of congestion

Congestion is due to insufficient road capacity (*recurrent congestion*, typically predictable and frequent) or other external causes (*non-recurrent congestion*, typically unpredictable and infrequent), such as inclement weather or incidents, of which incidents impact the traffic most [39, 40]. *Incident rates* are defined as number of incidents per million vehicle-km travelled [I/MVkmT]; therefore, the expected number of incidents strongly depends on the flow. In the context of bridge loading, the cause of an incident is not relevant; instead, its effects on the traffic capacity and subsequent congestion are relevant. For instance, the Highway Capacity Manual [34] suggests that a lane closure of a two-lane motorway drops the overall capacity by 65%. However, it must be noted that a capacity reduction does not necessarily cause congestion, as this would depend on the inflow Q_{in}.

Table 2 presents data about incidents from selected literature. Since long-span bridges typically have two lanes in the same direction, incidents reported to cause the closure of two (or more) lanes are here considered to fully block the road; hence, the corresponding rate is named *full-stop rate* FS_r (FS/MVkmT) [38]. Remarkably, while incident rates are spread over a wide range, full-stop rates cover a much smaller range.[5]

Acronym	Traffic state
FT	Free Traffic
MLC	Moving Localised Cluster
PLC	Pinned Localised Cluster
SGW	Stop-and-Go Waves
OCT	Oscillating Congested Traffic
HCT	Homogeneous Congested Traffic
FS	Full Stop

Table 1. Traffic states.

Author	Incident rate (I/MVkmT)	Incidents blocking 1 lane (%)	Incidents blocking ≥ 2 lanes (%)	Full-stop rate (FS/MVkmT)
Giuliano [41]	≈6	6.5	2	0.12
Skabardonis et al. [42]	64.6	3.1	0.6	0.39
Skabardonis et al. [43]	57.7	9.15	0.45	0.26
Rodgers et al. [44]	2.32	23.8	7.6	0.18
Tasnim et al. [45]	≈5	33	5	0.25

Table 2. Incident rates from selected literature and associated full-stop rates.

2.4. Microsimulation

The concept of modelling individual vehicles, namely *microsimulation*, is now well established for traffic studies [24, 46, 47]. Microsimulation takes account of the interaction between vehicles, as opposed to *macrosimulation*, which treats traffic as an aggregate flow. As traffic microsimulation is able to reproduce realistic spatial distributions of vehicles, it is a suitable tool to investigate load effects on bridges, without resorting to conservative assumptions about heavy-vehicle positions. Notably, widely available free-flowing traffic measurements can be used to generate initial and boundary traffic conditions. Congested data may be used to calibrate and validate the microsimulation parameters [48].

[5] This may be due to the fact that many small incidents can be unnoticed, while it is unlikely for a large incident causing lane closures to go unrecorded.

Many microsimulation models have been developed in the last few decades. The choice of a suitable microsimulation model mainly relies on the traffic features of interest, e.g. during free-flowing traffic or congestion. For bridge-loading applications, the microsimulation model should be able to reproduce the range of traffic states likely to occur on a bridge. Once calibrated, microsimulation enables the modelling of a large number of congestion events (and the subsequent identification of extreme loading events), which would be extremely difficult to record in the real world.

Microsimulation models divide into *car-following* (single-lane) and *lane-changing* (multi-lane) models. While in short-span load models lane-changing manoeuvres can be neglected, they play an important role in the context of long-span bridge loading. In fact, when overtaking is allowed, the car-truck mix for congestion is likely to be different to that for free traffic, since car drivers do not feel comfortable following trucks As traffic slows down and therefore tend to overtake them [49]. This typically results in longer truck-only platoons in congested traffic than in free-flowing traffic.

2.4.1. The Intelligent Driver Model

The *Intelligent Driver Model* (IDM) is a car-following model which has been shown to successfully replicate observed multi-lane congestion patterns on several motorways With few parameters using a single-lane traffic stream made up of identical vehicles with deterministic parameters [36, 50]. The IDM has also been proven able to replicate observed single-vehicle trajectory data [51–55]. The motion of each vehicle is simulated through an acceleration function:

$$\frac{dv(t)}{dt} = a\left[1-\left(\frac{v(t)}{v_0}\right)^4 - \left(\frac{s^*(t)}{s(t)}\right)^2\right] \tag{4}$$

in which a is termed the *maximum acceleration*, v_0 the *desired speed*, $v(t)$ the current speed, $s(t)$ the current gap to the front vehicle and $s^*(t)$ the *desired minimum gap*, given by:

$$s^*(t) = s_0 + \max\left\{Tv(t) + \frac{v(t)\Delta v(t)}{2\sqrt{ab}};0\right\} \tag{5}$$

in which s_0 is termed the *minimum jam* (bumper-to-bumper) *distance*, T the *safe time headway*, $\Delta v(t)$ the speed difference between the current vehicle and the vehicle in front and b the *comfortable deceleration*.[6] There are five parameters in this model (v_0, s_0, T, a, b) to capture driver behaviour, which are relatively easy to measure. For simulation purposes, the length of the

[6] The desired minimum gap s^* is limited to be greater than or equal to the minimum jam distance s_0, as noted in Eq. (5); otherwise, an inconsistent driver behaviour in multi-lane scenarios may be generated when the front vehicle is faster [18, 56].

http://dx.doi.org/10.5772/63967

vehicles must also be known. The congested states in **Table 1** can be effectively generated by applying an *inhomogeneity*, for instance, by increasing the safe time headway, T, downstream to – say – T'. For a comprehensive discussion, the reader is referred to [18] or [36].

Here, it is worth mentioning that for bridge-loading applications, the parameters T and s_0 greatly regulate the distance a vehicle keeps when following its leader, such that the smaller those parameters, the closer the vehicles and the Greater the loading. As speed tends to zero, the influence of the safe time headway T decreases and the spacing tends to the minimum jam distance s_0. Therefore, s_0 is a crucial parameter for bridge-loading applications [57].

The desired speed, v_0, regulates the behaviour in free traffic, Whereas the traffic stability[7] is mainly determined by a, b and T. Finally, it must be noted that the adoption of variable parameters among vehicles is not strictly necessary for reproducing the congested patterns [36, 58]. Further details about the application of the IDM to bridge-loading analysis can be found in [38].

2.4.2. *The MOBIL lane-changing model*

The MOBIL lane-changing model has been proposed in [59], to which the reader is referred For a detailed description. An overview of the model is given here, whereas details about the application of MOBIL to bridge loading can be found in [60].

Figure 3 depicts a lane change event, in which the subscript c refers to the lane-changing vehicle, o refers to the old follower (in the current lane) and n to the new one (in the target lane). All the accelerations, current and proposed (i.e. before and after the lane change), can be calculated according to the IDM given in Eqs. (4) and (5).

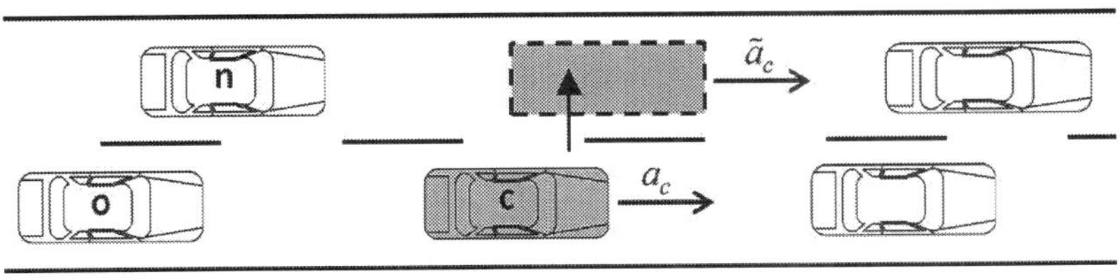

Figure 3. Vehicles involved in a lane-changing manoeuvre (adapted with permission from [59]).

A lane change occurs if both *incentive* and *safety* criteria are fulfilled. For a *slow-to-fast* lane change, the incentive criterion is expressed as follows:

[7] The traffic stability is the response of the traffic flow to a perturbation, e.g. a braking vehicle. If the traffic flow is unstable, then a perturbation will break the flow down and propagate into a stop-and-go wave (such as those shown in **Figure 1**).

$$\widetilde{a_c}(t) - a_c(t) > \Delta a_{th} + \Delta a_{bias} + p\left(a_n(t) - \widetilde{a_n}(t)\right) \tag{6}$$

This means that the acceleration advantage $\widetilde{a_c} - a_c$ in performing a lane change must be greater than the sum of the *acceleration threshold* Δa_{th}, which prevents overtaking with a marginal advantage, the *bias acceleration* Δa_{bias}, which acts as an incentive to keep in the slow lane, and the imposed disadvantage to the new follower in the fast lane $a_n - \tilde{a}_n$, weighted through a *politeness factor p*, to account for the driver aggressiveness. On the other hand, the incentive criterion for a *fast-to-slow* lane change is as follows:

$$\widetilde{a_c}(t) - a_c(t) > \Delta a_{th} - \Delta a_{bias} + p\left[\left(a_n(t) - \widetilde{a_n}(t)\right) + \left(a_o(t) - \widetilde{a_o}(t)\right)\right] \tag{7}$$

In this case, the acceleration advantage $\widetilde{a_c} - a_c$ must be greater than the sum of the acceleration threshold Δa_{th}, minus the bias acceleration Δa_{bias} (which acts as an incentive to move back to the slow lane), plus the disadvantage imposed to both new follower n in the slow lane and to the current follower o in the fast lane, weighted through the politeness factor p.[8]

Finally, the safety criterion limits the imposed deceleration to the follower n in the target lane to the *safe braking* value b_{safe}:

$$\tilde{a}_n(t) \geq -b_{safe} \tag{8}$$

2.5. Summary

Traffic data for bridge-loading applications is typically collected at high-speed WIM stations. Free-flowing traffic measurements are unbiased, reliable and now commonly available. Generally speaking, if traffic information is available at a point detector, such as a WIM station, it is possible to accurately reconstruct vehicle positions from single-vehicle data only when traffic characteristics do not change significantly. This is the usual case in free-flowing traffic and therefore applicable to short-span bridges.

On the contrary, there is a shortage of data during congestion, mainly Due to current technological limitations. In addition, the analysis of traffic data can pose some issues: in fact, a vehicle's speed is likely to fluctuate, e.g. as a result of *stop-and-go waves*, making the estimation of vehicle positions from point measurements problematic. The use of spatial detectors (such as cameras) over a stretch of road allows the collection of vehicle positions during congestion, without resorting to estimation. Although cameras are the best solution from a theoretical point of view, they are not often used for several practical reasons.

---

[8] Eq. (7) is based on [61] and preferred to the formulation in [59], which does not include the disadvantage to the new target follower $a_n - \tilde{a}_n$ in the fast-to-slow lane change.

Traffic microsimulation provides a valuable tool for long-span bridge loading, as it is capable of reproducing realistic congested scenarios from free-flowing traffic measurements. Load effects can be computed directly from the actual spatial distribution of vehicles, thus avoiding any inaccuracy due to the estimation of vehicle positions from point measurements.

3. Statistics concepts

In bridge loading, extreme events with small frequencies of occurrence are of interest, rather than frequent scenarios. However, it is not generally practicable to simulate the long periods required to identify such rare events, even when the recorded database is expanded through Monte Carlo techniques. Therefore, the data is typically fitted with a statistical distribution and then extrapolated to determine characteristic loading values.

While pioneering studies focussed on wors-case scenarios [5, 6], a probabilistic approach to bridge loading is now common [7–11, 27, 33, 38, 60, 62–66]. The probability F that a load level z is not exceeded (*probability of non-exceedance*) is commonly expressed in terms of *return period*, $T(z)$ [67]. The two variables are linked through the relation:

$$T(z) = \frac{1}{1 - F(z)} \tag{9}$$

Importantly, the return period is different from the bridge lifetime and, instead, should be seen as a measure of safety. For instance, the *characteristic* and *frequent* values of the Load Model 1 in Eurocode 1 are based on return periods of 1000 years and 1 week, respectively [68]. The superseded British HA loading was based on a return period of 2400 years [69]. The AASHTO load model is based on 75 years [70, 71]. For assessment, it is accepted that lower return periods should be used: in Europe, a value of 75 years may be considered, whereas AASHTO [72] suggests a return period of 5 years. As will be shown in Section 5.1, such large differences in the return period do not imply equally large differences in the extrapolated characteristic values.

3.1. Extreme value statistics

Extreme value theory is a branch of statistics Appropriate for the probabilistic modelling of a range of civil engineering problems. including bridge loading. A popular approach is the *Block Maxima*: the maximum events in each block (e.g. maximum load effect per day) are selected as representative, then fitted with a statistical distribution, and finally extrapolated to determine characteristic values. While other methods are possible, the database sample has generally more influence on the results than the extrapolation method, thus highlighting the importance of long-run simulations [73].

Firstly, an empirical frequency to each sample data point, z_i (*plotting position*), is assigned:

$$\hat{F}(z_i) = \frac{i}{N+1} \tag{10}$$

in which $i = 1, 2, \ldots, N$ is the index of the sample ordered decreasingly.[9] The *Generalised Extreme Value* (GEV) distribution is then fitted to the simulated maxima. Its *Cumulative Distribution Function* (CDF), which expresses the probability of non-exceedance $F(z)$, is as follows [67]:

$$F(z) = \exp\left\{-\left[1+\xi\left(\frac{z-\mu}{\sigma}\right)\right]^{-\frac{1}{\xi}}\right\} \tag{11}$$

in which μ is the location, σ the scale and ξ the shape parameter. Eq. (11) is defined for any value z for which $1 + \xi\left(\frac{z-\mu}{\sigma}\right) > 0$. When $\xi = 0$, the GEV distribution reduces to the Gumbel (or *Type I*) distribution:

$$F(z) = \exp\left\{-\exp\left[-\left(\frac{z-\mu}{\sigma}\right)\right]\right\} \tag{12}$$

When $\xi > 0$ and $\xi < 0$, the GEV distribution is named, respectively, Fréchet (or *Type II*) and Weibull (or *Type III*). The latter is more commonly found in bridge-loading applications. The GEV parameters can be inferred through *maximum likelihood estimation* (details can be found in [67]).

Gumbel probability paper plots are useful to illustrate the extrapolation procedure [74]. On this plot, data from a Gumbel distribution appears as a straight line. The ordinate, or *Standard Extremal Variate* (SEV), is given by:

$$SEV(z) = -log\left[-log\left(F(z)\right)\right] \tag{13}$$

Table 3 reports the target probabilities of exceedance and SEVs for typical return periods, under the common assumption that maxima are collected every day of a year with 250 working days.

3.2. The law of total probability

The GEV distribution applies under the assumption that individual events are independent and identically distributed. However, this assumption is not necessarily met in bridge-loading

[9] Alternative plotting position formulae are possible, but differences become small as the sample size increases.

applications: load effects can be the result of a number of quite different loading events, involving different numbers of trucks or different traffic states.

A generalisation of the *law of total probability* can be used to combine maximum load events resulting from different event types. The probability, P, that the combined maximum load effect, z, in a given reference period (e.g. a day or an hour) is not exceeded, i.e. the combined CDF, is:

$$P(z) = \sum_{j=1}^{n_t} F_j(z) \cdot f_j \tag{14}$$

in which n_t is the number of event types, F_j is the CDF of the maximum load effects for the j-event type (Eq. 11) and f_j is the probability of occurrence of the j-event type. Clearly, $\sum_{j=1}^{n_t} f_j = 1$. Equating $P(z)$ to the target probability (e.g. as in **Table 3**) gives the characteristic combined load.

Return period (years)	Probability of non-exceedance	SEV
2400	0.999998	13.30
1000	0.999996	12.43
75	0.999947	9.84
5	0.999200	7.13
0.02	0.800000	1.50

Table 3. Typical target probabilities of exceedance and SEV.

Eq. (14) may be applied to the combination of maximum load effects resulting from different j traffic states, each of which occurs with the assigned probability f_j [38]. It implies that, within the reference period, only *one* maximum loading event, z, can occur due to any of the j traffic states.

On the other hand, when considering load effects deriving from different j-truck meeting events, relevant to short-span bridge loading, Eq. (14) cannot be readily applied [75]. In fact, within any reference period, there will be n_t maximum loading events, each of which due to a j-truck meeting event. In this case, and only when using the GEV distribution (Eq. 11), it can be demonstrated that the probability, P, that the combined maximum load effect, z, is not exceeded in the reference period is given by [75]:

$$P(z) = \prod_{j=1}^{n_t} F_j(z) \tag{15}$$

in which n_t is the number of event types (typically 4) and F_j is the CDF of the GEV distribution for the maximum load effects of the j-event type.

4. Short-span bridges

Short-span bridges are governed by free-flowing traffic, plus an allowance for dynamic effects. Free-flowing traffic measurements can be used directly or as a basis to generate loading scenarios. The arrival of vehicles in free traffic is often idealised as a Poisson process, which can be described with a negative exponential distribution. If cars are neglected, as usually assumed in short-span bridge loading, other distributions may be more suitable to describe the arrival of truck traffic [76].

Firstly, the case of an individual lane is considered. Shorter bridge spans (< ≈30 m) are governed by a single heavy vehicle or, if very short, individual axles. Therefore, commonly available information about axle configuration and weight suffices to generate a realistic (static) load model, as there is no need to account for vehicle interaction.

For longer spans, in-lane multiple presence of heavy vehicles is a possible event. The headway between two following trucks is an important parameter to identify the number of trucks that may be simultaneously present on a bridge. Headways may be estimated from the WIM database, as discussed in Section 2.2, or from a calibrated headway model [76].

For the common case of multi-lane bridges, it is necessary to consider the multiple presence of Side-by-side heavy vehicles, whether across same-direction or opposing lanes. In the development of current European and North American codes, data about actual multiple presence of side-by-side vehicles was not collected but artificially reproduced. For instance, Reference [66] considers that one in 15 trucks has another truck side by side. This conservative assumption, used for the calibration of the current AASHTO load model [71], can be nowadays adapted to site-specific traffic conditions due to the availability of more accurate WIM data [77, 78]. Indeed, a 2-truck side-by-side meeting event is likely to strongly influence the bridge design for shorter spans (< ≈30 m) and sometimes it has also been the only considered event for longer spans [66, 79]. However, as spans get longer, the likelihood of events involving more than two trucks increases and such meeting events should be accounted for [75, 76, 80].

4.1. Dynamic amplification factor

Fast-moving vehicles, typical of free-flowing traffic, interact dynamically with a bridge. The total load effect resulting from a loading event, LE_t, is typically larger than would result from a static analysis, LE_s. The *Dynamic Amplification Factor* (DAF) represents the ratio of these two load effects:

$$DAF = \frac{LE_t}{LE_s} \tag{16}$$

Dynamic amplification varies significantly and depends on a number of factors, such as span length, vehicle class and speed, axle spacing and weight, suspension stiffness, or the road surface profile. As described in the next section, the main codes use DAFs in the range from 1.0 to 1.8.

Reference [81] presents a review about dynamic factors for highway bridges. In recent years, the estimation of dynamic allowance has shifted from a worst-case scenario point-of-view to a probabilistic approach. In fact, several studies have identified the potential for a large DAF reduction when the static traffic loading increases, as the maximum dynamic effect does not correspond to the maximum static effect. For instance, for a single-vehicle event, there is a significant probability that a vehicle will travel at a speed which excites the bridge, whereas for a two-vehicle meeting event it is much less likely that both vehicles contribute to dynamic amplification. Moreover, a heavier vehicle typically excites the bridge less than a lighter one.

To address this issue, the *Assessment Dynamic Ratio* (ADR) has been proposed as the ratio between the *total characteristic* load effect and the *static characteristic* load effect [2]. These two characteristic values may not necessarily arise from the same loading scenario. Numerical investigations on dynamic vehicle-bridge interaction have found that the ADR is in the order of 1.05 [2, 82]. This considerable difference from typical dynamic allowances has the impact that many longer bridges thought to be governed by free-flowing traffic could actually be governed by congestion conditions [4]. Therefore, the simulation of congested traffic, described in Section 5.1, may have a wider application to That originally thought.

4.2. Main codes

The Load Model 1 (LM1) in Eurocode 1 [68] was calibrated with 1-week WIM data collected on the A6 Motorway near Auxerre (France). The traffic scenarios were based on a dual two-lane carriageway layout. In the simulated free-flowing traffic (spans up to 50 m) a maximum of 25% trucks was considered. Nine influence lines were investigated for nine spans ranging from 5 to 200 m [64]. Several extrapolation methods were tested to find the characteristic values corresponding to the target return period of 1000 years; in general, results were not found to be very sensitive to the extrapolation technique [69]. LM1 is given as a tandem axle and a Uniformly Distributed Load (UDL), whose values depend on the lane but not on the span length. LM 1 incorporates dynamic effects: the considered DAFs were in the range 1.0 to 1.7, with the greater value for shorter span [64]. LM1 may be significantly reduced when used for site-specific bridge assessment [83].

The British code for the design of bridges [84] has been now superseded by the Eurocode. However, its main load model – The HA loading – is still prescribed for the assessment of existing highway bridges up to 50 m [85]. The HA loading is based on UK legal limits and data from roadside truck surveys [69] and consists of a UDL, given as a loading curve depending on the span length, and a Knife Edge Load (KEL). Multi-lane factors are given to consider further lanes. A DAF of 1.8 is included for single-vehicle cases [69]. A *Reduction Factor* between 0.2 and 0.91 is applied to the HA design loading depending on the required assessment level, heavy traffic proportion and road surface condition [69, 85].

In North America, the HL-93 model in AASHTO [71] is based on truck surveys taken in Ontario and weigh-in-motion data. Calibration has been carried out for one- and two-lane girder bridges with single and two spans from 9 to 60 m [70, 86]. The model consists of a design truck, or a tandem, coincident with a design lane load. Multi-lane factors are also specified. It is specified to apply a DAF of 1.33, independent of length or load effect.

5. Long-span bridges

Long-span bridges are governed by congested conditions, with no allowances for dynamic effects, due to the slow speed of the vehicles involved in critical loading events. Traditionally, long-span bridge loading has been based on the simulation of queues of vehicles [5–11, 27, 63, 64]. As outlined in Section 2.4, traffic microsimulation is a powerful tool to simulate more realistic congested scenarios for long-span bridge loading.

5.1. Application of microsimulation to bridge loading

An application of traffic microsimulation to long-span bridge loading is presented for a stretch of a two-lane same-direction 8000 m long highway, based on [60]. The microsimulation is carried out using the car-following IDM (Section 2.4.1) and the lane-changing model MOBIL (Section 2.4.2). In order to highlight the influence of several traffic features on bridge loading, a simplified vehicle stream made up of two classes of vehicle, cars and trucks, is used with the parameters shown in **Table 4**. Real sites are likely to have a more complex traffic stream, but site-specific traffic data could be equally introduced [87].

The car-following parameters are based on those calibrated and used in [36], who used only identical vehicles to successfully replicate obseved congestion patterns, as described in Section 2.4.1. Trucks are introduced here and assigned greater length and weight and slower desired speed [88, 89]. Other truck parameters are kept the same as the parameter set in [36], as consideration of different parameters is not strictly necessary to reproduce congested patterns [36, 58].

The desired speeds, v_0, of both vehicle classes are uniformly randomly distributed. Although the desired speed governs the free traffic behaviour in the IDM, it is necessary to introduce speed randomness in order to correctly model lane-changing manoeuvres [59]. Trucks are assigned mean Gross Vehicle Weight (GVW) equal to the European minimum legal limit of 44 t [90, 91] and the GVW is considered normally distributed with a Coefficient of Variation (CoV: standard deviation divided by the mean) of 10%. Those two assumptions can be easily adapted to a specific site, as truck GVWs and speed histograms can be computed from WIM.

Cars and trucks are assigned the same MOBIL parameters, as calibrated in [15]. However, the assigned difference in the IDM desired speed, v_0, has also the desirable effect of making trucks less prone to undertake a lane-changing manoeuvre. Notably, bridge loading is not found to be significantly sensitive to the lane-changing parameters [15].

Two long-span bridges (200 and 1000 m long) are centred at 5000 m. The dynamic capacity Q_{out} is 3070 veh/h for a flow with 20% trucks. A range of bottleneck strengths ΔQ is generated downstream of the bridge by locally increasing the safe time headway T from 1.6 s to the values of 1.9, 2.2, 2.8, 4.0 and 6.4 s, thereby inducing the following traffic states: SGW, OCT, HCT/OCT, HCT(1) and HCT(2). HCT(1) and HCT(2) differ for the average speed of traffic (approximately 8.7 and 5.0 km/h). The full-stop condition (FS) is also simulated, for which $\Delta Q = Q'_{out}$.

It is assumed that congestion occurs for one hour during every working day (i.e. 250 times per year). One-year equivalent of traffic is simulated for each bottleneck strength. The hourly/daily maxima of total load are captured for subsequent statistical extrapolation with the Block Maxima approach, as described in Section 3.1. Note that load effects, such as shear or bending moment, may be equally output. However, this requires the choice of a structural form for the bridges. Since such forms may be quite different for long spans, the total load is used here to maintain generality.

5.1.1. *Identification of critical traffic states*

The hourly maxima of total load for a flow of 3000 veh/h with 20% trucks on the two spans are plotted in **Figure 4**, along with the extrapolated characteristic values corresponding to return periods of 5, 75 and 1000 years.

	Cars	Trucks
Desired speed, v_0	120 km/h (±20%)	80 km/h (±10%)
Safe time headway, T	1.6 s	1.6 s
Maximum acceleration, a	0.73 m/s^2	0.73 m/s^2
Comfortable deceleration, b	1.67 m/s^2	1.67 m/s^2
Minimum jam distance, s_0	2 m	2 m
Vehicle length, l	4 m	12 m
Politeness factor, p	0.1	0.1
Changing threshold, Δa_{th}	0.2 m/s^2	0.2 m/s^2
Bias for the slow lane, Δa_{bias}	0.2 m/s^2	0.2 m/s^2
Maximum safe deceleration, b_{safe}	6 m/s^2	6 m/s^2
Gross vehicle weight	20 kN	432 kN*

* Coefficient of variation 0.1.

Table 4. Model parameters.

In order to consider the case of the combination of congestions (Eq. 14), it is necessary to assign the congestion frequencies f_j [38]. Firstly, the number of expected full-stop events, FS_n is computed as follows:

$$FS_n = FS_r \frac{ADT \cdot L \cdot T}{10^6} \tag{17}$$

in which FS_r is the full-stop rate (**Table 2**), ADT the Average Daily Traffic (veh/day), L the length of road observed (km) and T the duration of observations (days). Here it is assumed that FS_r

= 0.25 FS/MVkmT [45], ADT = 32000 veh/day, T = 250 days (equivalent to one year) and L = 5 km (i.e. incidents occurring up to 5 km downstream of the bridge will affect it), thus returning 10 expected full-stop events each year. Following the previous assumption of 250 congestion events per year, this corresponds to a frequency of 4% over the total number of congestion events.

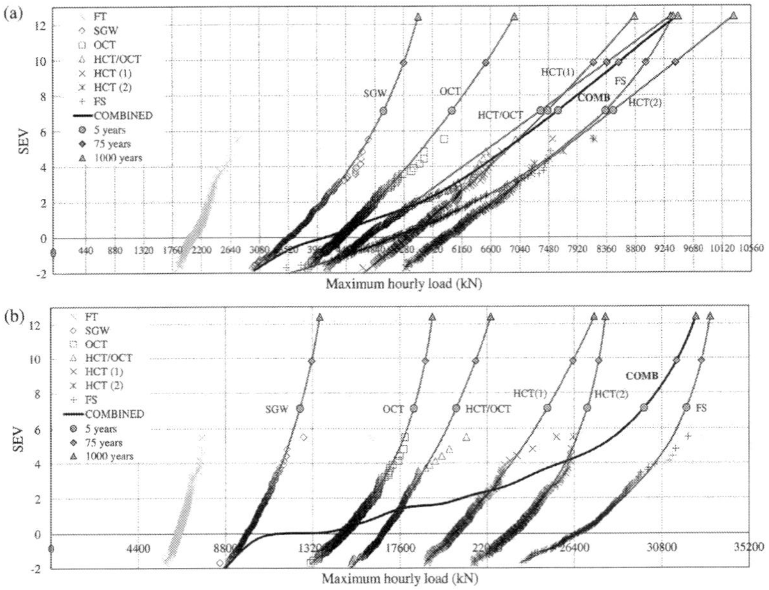

Figure 4. Probability paper plots of maximum total load: (a) 200 m span; (b) 1000 m span.

The expected frequency of full-stop events is used as the basis for assigning congestion frequencies to the other bottleneck strengths generated (**Figure 5**). The distribution is taken to be exponential [38] and qualitatively agrees with available observations [37].

Several features can be noticed from the probability paper plots of **Figure 4**:

- Free traffic maxima are much lower than the congested ones.[10]
- Full stop is not the most critical traffic state for the 200 m span (**Figure 4a**) . In fact, slow-moving HCT states give more combinations of vehicles (and subsequently more chances of finding an extreme loading scenario), whereas FS is the maximum of only one realisation of vehicles, although vehicles are spaced at the minimum jam distance s_0.
- For the 1000 m span (**Figure 4b**), FS is the most critical condition even at small return periods.

[10] Free traffic maxima are shown only for reference and have not been included in the statistical analysis.

- Full-stop CDFs are characterised with a greater curvature (smaller ξ) and variability (greater σ) (**Table 5**).
- Oscillating congested states (SGW and OCT) do not govern the combined load, other than for small SEVs (return period less than about a week). They may actually be disregarded, as done with FT, with negligible effects on the combined characteristic load corresponding to typical return periods.[11]
- Separation between congested states is sharper for the 1000 m span, since local concentrations of heavy vehicles are averaged out as span increases.
- For the 200 m span (**Figure 4a**), considering only the most adverse congested state (HCT(2)) occurring every working day overestimates the total load by about 11% at 5 years and 9% at 1000 years, suggesting that the error in assuming only the worst congested state slightly reduces as return period increases.
- For the 1000 m span (**Figure 4b**), considering only the most adverse congested state (FS) occurring every working day overestimates the total load by about 7% at 5 years and 2% at 1000 years. Hence, the error in assuming only the worst congested state decreases further with increasing span.

Finally, it is noted that traffic microsimulation may be applied to free traffic as well. However, its computational requirements make this approach less attractive, compared to simpler models. Moreover, the IDM shows an excessive spreading of platoons when vehicles approach the desired speed v_0 [15]. This has essentially no effect on the IDM capability of reproducing

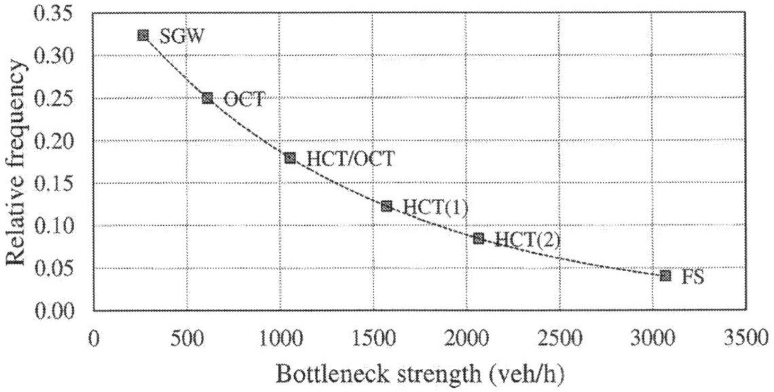

Figure 5. Congestion frequencies.

[11] It can be deduced from **Figure 6** that SGW and OCT occur in 57.3% of the working day. If the analysis is carried out neglecting SGW and OCT, the critical congested states HCT/OCT, HCT and FS occur on 106.7 working days per year. It can be shown that re-running the procedure above with adjusted congestion frequencies and target probabilities of non-exceedance returns nearly identical characteristic load values.

congested traffic states (as $v << v_0$), but could potentially affect bridge loading in uncongested situations. To avoid this issue, a suitable modification of the IDM has been proposed [92].

Span	**200 m**			**1000 m**		
Congested state	HCT(1)	HCT(2)	FS	HCT(1)	HCT(2)	FS
Location parameter, μ	5340	5908	5140	20334	23212	26519
Scale parameter, σ	328.7	366.4	694.5	811.1	954.0	1415.3
Shape parameter, ξ	−0.028	−0.006	−0.132	−0.061	−0.179	−0.190

Table 5. Parameters of the GEV distribution for HCT and FS conditions.

5.1.2. *Influence of some traffic features*

Figure 6 shows the 5-year characteristic values for the inflows of 3000 (as described in Section 5.1.1), 2000 and 1250 veh/h, expressed as an Equivalent Uniformly Distributed Load (EUDL, total load divided by bridge length). EUDLs resulting from a flow of 1250 veh/h with 48% trucks are also plotted, in order to quantify the influence of truck percentage and car presence on loading (the 1250 veh/h flow with 48% trucks has the same truck flow of the 3000 veh/h flow with 20% trucks).[12]

In general, the loads resulting from different inflows with same truck percentage and same bottleneck strength are quite similar (within a range of ±11%), as long as the bottleneck is strong enough to trigger congestion for that inflow, that is $Q_{in} > Q'_{out}$. This suggests that the effect of the inflow on loading is not as strong as that of bottleneck strength and implies that critical loading events may occur also out of rush hours.[13]

Among the traffic features affecting bridge loading, the truck percentage is of particular significance. High truck percentages may occur at night time and early morning in correspondence to low overall flows. As such, it is rare to have congestion events, but if they occur, the resulting loading is likely to be quite heavy, as there are not many cars to keep heavy vehicles apart.

Figure 6 also shows that at lower flows (1250 veh/h) FS governs for both spans and that the influence of truck percentage on the total load is expectedly large: +27.5% and +40.2% for FS, respectively, on the 200 and 1000 m span. On the other hand, the presence of cars reduces the loading by about 12% (200 m) and 29% (1000 m). Therefore, it is advised that traffic data collection should be based also on periods of low flow but with high truck percentage.

Finally, it is worth noting that the assignment of realistic congestion frequencies is certainly important for a correct bridge-loading analysis. As mentioned in Section 2.3.1, some sites are

[12] Note that the increased truck percentage reduces the dynamic capacity Q_{out}, as mentioned in Section 2.3. Therefore, it is necessary to apply a greater modified safe time headway $T' = 6.0$ s to reproduce a HCT(2) state similar to that of a flow with 20% trucks [60].

[13] An exception is the HCT(2) state at 1250 veh/h for the 1000 m span. However, this is due to the slower congestion growth occurring at lower flows, which takes somewhat less than an hour to fill the bridge up, thus giving fewer realizations of critical events.

likely to be congested on a daily basis (e.g. urban corridors), whereas others can hardly experience any congestion (e.g. intercity motorways). Fortunately, the characteristic load is not largely sensitive to the congestion frequency [9, 15, 38]: it would generally suffice to consider a reasonable estimate. For instance, for the data reported above, halving the number of congestion events to 125 per year decreases the 5-year SEV to 6.44 (Eq. 13), thereby reducing the corresponding characteristic loading for the 200 m and the 1000 m respectively by 3.4% and 2.0%. If the expected congestion events are ten times less, which could make the difference between sites affected by recurrent or non-recurrent congestion, the 5-year SEV is 4.82; in this case, reductions become more significant (12.4% and 9.1%). Note that in the latter case, it is an *interpolation* of the simulated data that has been carried out.

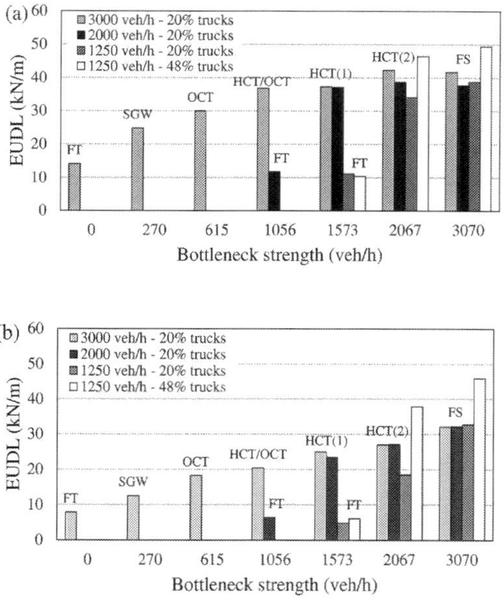

Figure 6. 5-year EUDL for: (a) 200 m span; (b) 1000 m span.

5.2. Main codes

The calibration studies of the Eurocode LM1 considered for jammed traffic (spans 75-200 m) a queue of heavy vehicles spaced at 5 m (axle to axle) in the slow lane [64]. National Annexes may extend the span limit of application of LM1. For instance, the UK National Annex extends its application up to 1500 m. Again, LM1 may be significantly reduced when using site-specific traffic data, with traffic microsimulation used to replicate congestion events [87].

The superseded British standard [84], which specified long-span bridge loading in a similar form to the HA loading for short spans (UDL + KEL), was greatly based on [9]: truck surveys and free-flowing traffic data were used to build up queues of heavy vehicles and cars, with gaps varying from 0.9 to 1.8 m; a simple modelling of lane choice of vehicles approaching a queue was also considered.

In North America, the American Society of Civil Engineers (ASCE) recommended a load model for the design of spans up to 1951 m [93]. A UDL is to be applied in conjunction with a point load, whose values depend on span length and truck percentage. The ASCE loading is mainly based on truck data from crossings of the Second Vancouver Narrows bridge [7]. Notably, 800 full-stop events per year were considered with vehicles spaced at 1.5 m.

6. Summary and conclusions

In this chapter, traditional approaches and recent advances in highway bridge traffic loading are described. These are of great significance for structural safety assessment of bridges, where there is a potential for substantial savings by considering site-specific traffic conditions. An introduction to traffic theory and modelling relevant to bridge-loading applications is given, as well as an overview of extreme value statistics, since a probabilistic approach is now well established when studying bridge loading.

In bridge traffic loading, it is convenient to distinguish between short-span bridges, which are governed by free-flowing traffic plus an allowance for dynamic vehicle-bridge interaction, and long-span bridges, which are governed by congested conditions with no allowance for dynamic effects. The bridge length threshold between the two modes is not clear-cut but is thought to lie between 30 and 50 m.

Current technologies allow the collection of a great deal of traffic data during uncongested traffic conditions, mainly from *weigh-in-motion* stations. Such data can be used for the analysis of short-span bridges. Importantly, recent studies have shown that dynamic allowances may be significantly smaller than those considered in the main codes of practice, especially when favourable site-specific conditions are accounted for. This implies that the threshold between short- and long-span bridges may be lower than currently thought and that recent techniques to simulate congested traffic, such as those described in this chapter, may have a wider application than expected.

On the other hand, a shortage of suitable congested data has led to the fact that traffic loading on long-span bridges is often based on conservative assumptions, traditionally a queue of vehicles at minimum bumper-to-bumper distances. Traffic *microsimulation* is a powerful tool to generate realistic congestion patterns based on the widely available free-flowing traffic measurements. Among microsimulation models, the *Intelligent Driver Model* provides an optimal balance between accuracy and computational speed and can be extended with the lane-changing model MOBIL to simulate the remixing of cars and trucks occurring as traffic gets congested. Calibration of the model parameters can be based on site-specific traffic data or on available data in the literature.

Here, it is shown that microsimulation can be effectively integrated into traditional structural analysis techniques to study the effect of different traffic features on bridge loading and compute a site-specific traffic loading. Simulations on two sample spans (200 and 1000 m long) show that, besides full-stop conditions, slow-moving Homogeneous Congested Traffic (HCT) can be critical. Among several traffic features analysed, the bottleneck strength and truck percentage are found predominant. In comparison, overall traffic flow or truck traffic flow do not significantly affect total loading. Therefore, traffic data collection should also focus on periods characterised by a high truck percentage, likely to occur at night time or early morning.

In conclusion, improved computer performances are likely to make a microsimulation-based approach to highway bridge traffic loading increasingly attractive. A site-specific traffic loading can be then computed, thus allowing a more efficient planning of costly maintenance operations.

Acknowledgements

This chapter draws several findings from the research conducted by the author at University College Dublin under the supervision of Prof E. OBrien and Dr C. Caprani within the TEAM (*Training in European Asset Management*) project and funded by the European Commission 7th Framework Programme. The author is most grateful to all contributors to that project and also acknowledges the valuable comments of Dr M. Treiber at Technische Universität Dresden on Section 2.

Author details

Alessandro Lipari

Address all correspondence to: alessandrolipari.ucd@gmail.com

TSP Projects, Meridian House, The Crescent, York, United Kingdom

References

[1] Moses F. Calibration of load factors for LRFR bridge evaluation. Washington: National Cooperative Highway Research Program; 2001, Report No.: 454.

[2] Caprani CC, González A, Rattigan PH, OBrien EJ. Assessment dynamic ratio for traffic loading on bridges. Structure and Infrastructure Engineering: Maintenance, Managemenet, Life-Cycle Design and Performance. 2012;8(3):295–304. DOI: 10.1080/15732471003667645

[3] González A, Rattigan PH, OBrien EJ, Caprani CC. Determination of bridge lifetime dynamic amplification factor using finite element analysis of critical loading scenarios. Engineering Structures. 2008;30(9):2330–2337. DOI: 10.1016/j.engstruct.2008.01.017

[4] Caprani CC. Lifetime highway bridge traffic load effect from a combination of traffic states allowing for dynamic amplification. Journal of Bridge Engineering. 2013;18(9): 901–909. DOI: 10.1061/(ASCE)BE.1943-5592.0000427

[5] Henderson W. British highway brigde loading. Proceedings of the Institution of Civil Engineers. 1952;3(3):325–350. DOI: 10.1680/ipeds.1954.11638

[6] Ivy RJ, Lin TY, Mitchell S, Raab NC, Richey VJ, Scheffey CF. Live loading for long-span highway bridges. American Society of Civil Engineers Transactions. 1954;119:981–1004.

[7] Buckland PG, Navin FPD, Zidek JV, McBryde JP. Proposed vehicle loading of long-span bridges. Journal of the Structural Division. 1980;106(4):915–32.

[8] Harman DJ, Davenport AG, Wong WSS. Traffic loads on medium and long span bridges. Canadian Journal of Civil Engineering. 1984;11(3):556–573.

[9] Flint and Neill Partnership. Interim design standard: Long span bridge loading. Crowthorne: Transport and Road Research Laboratory; 1986, Contract No.: Contractor Report 16.

[10] Ditlevsen O, Madsen HO. Stochastic vehicle-queue-load model for large bridges. Journal of Engineering Mechanics. 1994;120(9):1829–1847.

[11] Nowak AS, Lutomirska M, Sheikh Ibrahim FI. The development of live load for long span bridges. Bridge Structures. 2010;6(1):73–79. DOI: 10.3233/BRS-2010-006

[12] O'Connor C, Shaw PA. Bridge loads: An international perspective. London: Spon Press; 2000.

[13] Ditlevsen O, Madsen HO. Structural reliability methods 2005.

[14] Klein LA, Mills MK, Gibson DRP. Traffic detector handbook. MacLean, VA: Federal Higway Administration; 2006. Contract No.: FHWA-HRT-06-108.

[15] Lipari A. Micro-simulation modelling of traffic loading on long-span bridges [PhD thesis]. Dublin: University College Dublin; 2013.

[16] Wardrop JG. Some theoretical aspects of road traffic research. Proceedings of the Institution of Civil Engineers: Engineering Divisions. 1952;1(3):325–362.

[17] Hall FL. Traffic stream characteristics. In: Gartner N, Messer CJ, Rathi AK, editors. Revised Monograph on Traffic Flow Theory. Federal Highway Administration. Washington; 1994.

[18] Treiber M, Kesting A. Traffic flow dynamics: Data, models and simulation. Heidelberg. Springer; 2013. 503 p.

[19] Edie LC. Flow theories. In: Gazis DC, editor. Traffic science. New York: Wiley Interscience; 1974. p. 1–108.

[20] Cassidy MJ, Coifman B. Relation among average speed, flow and density. Transportation Research Record. 1997;1591:1–6.

[21] Treiterer J, Myers JA. The hysteresis phenomenon in traffic flow. In: Buckley DJ, editor. 6th International Symposium on Transportation and Traffic Theory; 1974; 26–28 August; Sydney: Elsevier.

[22] Hoogendoorn SP, van Zuylen HJ, Schreuder M, Gorte B, Vosselman G. Microscopic traffic data collection by remote sensing. Transportation Research Record: Journal of the Transportation Research Board. 2003;1855:121–128.

[23] Wilson RE. From inductance loops to vehicle trajectories. In: Beal GJ, editor. 75 years of the fundamental diagram for traffic flow theory. Washington: Transportation Research Board; 2011. p. 134–143.

[24] Orosz G, Wilson RE, Stépán G. Traffic jams: Dynamics and control. Philosophical Transactions of the Royal Society A. 2010;368:4455–4479. DOI: 10.1098/rsta.2010.0205

[25] Enright B, OBrien EJ. Monte Carlo simulation of extreme traffic loading on short and medium span bridges. Structure and Infrastructure Engineering: Maintenance, Management, Life-Cycle Design and Performance. 2013;9(12):1–16. DOI: 10.1080/15732479.2012.688753

[26] Eymard R, Jacob B. Un nouveau logiciel: A new software: the program CASTOR for the calculation of actions and effects due to traffic on highway structures. Bulletin de Liaison des LPC. 1989;164:64–77.

[27] Bailey SF, Bez R. Site-specific probability distribution of extreme traffic action effects. Probabilistic Engineering Mechnanics. Bridge and Concrete Research (BCRI), Ireland. 1999;14(1):19–26.

[28] Coifman B, Beymer D, McLauchlan P, Malik J. A real-time computer vision system for vehicle tracking and traffic surveillance. Transportation Research Part C: Emerging Technologies. 1998;6:271–288.

[29] Federal Highway Administration. NGSIM project. Federal Highway Administration; 2005 [cited 2012 02 February]; Available from: http://www.ops.fhwa.dot.gov/trafficanalysistools/ngsim.htm.

[30] Blacoe S, Caprani CC, OBrien EJ, Lipari A. Determination of minimum gap in congested traffic. In: Caprani CC, O'Connor AJ, editors. Bridge and Concrete Research in Ireland; 2012, 6-7 September; Dublin.

[31] Ricketts NJ, Page J. Traffic data for highway bridge loading. Crowthorne: Transport Research Laboratory; 1997, Contract No.: 251.

[32] Zaurin R, Catbas FN. Integration of computer imaging and sensor data for structural health monitoring of bridges. Smart Materials and Structures. 2010;19:1–15.

[33] OBrien EJ, Hayrapetova A, Walsh C. The use of micro-simulation for congested traffic load modelling of medium- and long-span bridges. Structure and Infrastructure Engineering: Maintenance, Management, Life-Cycle Design and Performance. 2012;8(3):269–276. DOI: 10.1080/15732471003640477

[34] Transportation Research Board. Highway capacity manual HCM 2010. Volume 2: Uninterrupted flow. Washington: Transportation Reseach Board; 2010.

[35] Cassidy MJ, Bertini RL. Some traffic features at freeway bottlenecks. Transportation Research Part B: Methodological. 1999;33:25–42.

[36] Treiber M, Hennecke A, Helbing D. Congested traffic states in empirical observations and microscopic simulations. Physical Review E. 2000;62(2):1805–1824.

[37] Schönhof M, Helbing D. Empirical features of congested traffic states and their implications for traffic modelling. Transportation Science. 2007;41(2):135–166. DOI: 10.1287/trsc.1070.0192

[38] OBrien EJ, Lipari A, Caprani CC. Micro-simulation of single-lane traffic to identify critical loading conditions for long-span bridges. Engineering Structures. 2015;94:137–148. DOI: 10.1016/j.engstruct.2015.02.019

[39] Kwon J, Mauch M, Varaiya P. Components of congestion: Delay from incidents, special events, lane closures, weather, potential ramp metering gain, and excess demand. Transportation Research Record: Journal of the Transportation Research Board. 2006;1959:84–91.

[40] Cambridge Systematics Inc. Traffic congestion and reliability: Trends and advanced strategies for congestion mitigation. Cambrigde: Cambridge Systematics Inc., 2005.

[41] Giuliano G. Incident characteristics, frequency, and duration on a high volume urban freeway. Transportation Research Part A: General. 1989;23(5):387–396.

[42] Skabardonis A, Petty KF, Bertini RL, Varaiya PP, Noeimi H, Rydzewski D. I-880 field experiment: Analysis of incident data. Transportation Research Record. 1997;1603:72–79.

[43] Skabardonis A, Petty KF, Varaiya PP. Los Angeles I-10 field experiment: Incident patterns. Transportation Research Record. 1999;1683:22–30.

[44] Rodgers S, Wadsworth B, Smith SD, Forde MC. An analysis of road traffic incidents on the M25 motorway, UK. Proceedings of the ICE - Transport. 2006;159(1):1–8.

[45] Tasnim S, Monsere CM, Bertini RL. Toward an automated incident analysis process using archived data on the Portland Oregon freeway system. 10th International Conference on Application of Advanced Technologies in Transportation. 2008; 27–31 May; Athens.

[46] Brackstone M, McDonald M. Car following: A historical review. Transportation Research Part F: Traffic Psychology and Behaviour. 1999;2:181–196.

[47] Hoogendoorn SP, Bovy PHL. State-of-the-art of vehicular traffic modelling. Proceedings of the Institution of Mechanical Engineers, Part I: Journal of Systems and Control Engineering. 2001;215:283–303.

[48] Hidas P, Wagner P. Review of data collection methods for microscopic traffic simulation. 10th World Conference on Transport Research; 2004 4–8 July; Istanbul.

[49] Peeta S, Zhang P, Zhou W. Behavior-based analysis of freeway car-truck interactions and related mitigation strategies. Transportation Research Part B: Methodological. 2005;39:417–451. DOI: 10.1016/j.trb.2004.06.002

[50] Helbing D, Treiber M, Kesting A, Schönhof M. Theoretical vs. empirical classification and prediction of congested traffic states. The European Physical Journal B. 2009;69:583–598. DOI: 10.1140/epjb/e2009-00140-5

[51] Kesting A, Treiber M. Calibrating car-following models by using trajectory data. Transportation Research Record: Journal of the Transportation Research Board. 2008;2088:148–156.

[52] Chen C, Li L, Hu J, Geng C. Calibration of MITSIM and IDM car-following model based on NGSIM trajectory datasets. International Conference on Vehicular Electronics and Safety. 2010; 15–17 July; QingDao (China).

[53] Hoogendoorn SP, Hoogendoorn R. Calibration of microscopic traffic-flow models using multiple data sources. Philosophical Transactions of the Royal Society A. 2010;368:4497–4517. DOI: 10.1098/rsta.2010.0189

[54] Brockfeld E, Kühne RD, Wagner P. Calibration and validation of microscopic traffic flow models. Transportation Research Record: Journal of the Transportation Research Board. 2004;1876:62–70.

[55] Punzo V, Simonelli F. Analysis and comparison of microscopic traffic flow models with real traffic microscopic data. Transportation Research Record: Journal of the Transportation Research Board. 2005;1934:53–63.

[56] Caprani CC, OBrien EJ, Lipari A. Extension of a lane-changing model to a microsimulation tool. Irish Transportation Research Network; 2011 31 August–1 September; Cork.

[57] Caprani CC. Calibration of a congestion load model for highway bridges using traffic microsimulation. Strucural Engineering International. 2012;22(3):342–348.

[58] Kesting A, Treiber M. How reaction time, update time, and adaptation time influence the stability of traffic flow. Computer-Aided Civil and Infrastructure Engineering. 2008;23:125–137.

[59] Kesting A, Treiber M, Helbing D. DOI: 10.3141/1999-10 General lane-changing model MOBIL for car-following models. Transportation Research Record: Journal of the Transportation Research Board. 2007;1999:86–94.

[60] Caprani CC, OBrien EJ, Lipari A. Long-span bridge traffic loading based on multi-lane traffic micro-simulation. Engineering Structures. 2016;115:207–19. DOI: 10.1016/j.engstruct.2016.01.045

[61] Treiber M. Microsimulation of road traffic flow. 2011 [accessed 2012 30 May]; Available from: http://www.traffic-simulation.de/.

[62] Harman DJ, Davenport AG. A statistical approach to traffic loading on highway bridges. Canadian Journal of Civil Engineering. 1979;6:494–513.

[63] Vrouwenvelder ACWM, Waarts PH. Traffic loads on bridges. Structural Engineering International. 1993;3/93:169–177.

[64] Prat M. Traffic load models for bridge design: Recent developments and research. Progress in Structural Engineering and Materials. 2001;3:326–334.

[65] Crespo-Minguillón C, Casas JR. A comprehensive traffic load model for bridge safety checking. Structural Safety. 1997;19(4):339–359.

[66] Nowak AS. Live load model for highway bridges. Structural Safety. 1993;13:53–66.

[67] Coles S. An introduction to statistical modeling of extreme values. London: Springer; 2001.

[68] European Committee for Standardization. Eurocode 1: Actions on structures. Part 2: Traffic loads on bridges. Brussels. CEN; 2003.

[69] Dawe P. Traffic loading on highway bridges. London: Thomas Telford; 2003. 168 p.

[70] Nowak AS. Calibration of LRFD bridge code. Journal of Structural Engineering. 1995;121(8):1245–1251.

[71] American Association of State Highway and Transportation Officials. AASHTO LRFD bridge design specifications, 6th edition. Washington: American Association of State Highway and Transportation Officials; 2012.

[72] American Association of State Highway and Transportation Officials. The manual for bridge evaluation, 2nd edition. Washington: American Association of State Highway and Transportation Officials; 2011.

[73] OBrien EJ, Schmidt F, Hajializadeh D, Zhou X-Y, Enright B, Caprani CC, et al. A review of probabilistic methods of assessment of load effects in bridges. Structural Safety. 2015;53:44-56. DOI: 10.1016/j.strusafe.2015.01.002

[74] Ang AH-S, Tang WH. 420 p. Probability concepts in engineering: Emphasis on application to civil and environmental engineering, 2nd edition. Hoboken: Wiley; 2007.

[75] Caprani CC, OBrien EJ, McLachlan GJ. Characteristic traffic load effects from a mixture of loading events on short to medium span bridges. Structural Safety. 2008;30:394–404. DOI: 10.1016/j.strusafe.2006.11.006

[76] OBrien EJ, Caprani CC. Headway modelling for traffic load assessment of short to medium span bridges. The Structural Engineer. 2005;83:33–36.

[77] OBrien EJ, Enright B. Modeling same-direction two-lane traffic for bridge loading. Structural Safety. 2011;33:296–304. DOI: 10.1016/j.strusafe.2011.04.004

[78] Sivakumar B, Ghosn M, Moses F, TranSystems Corporation. Protocols for using and collecting traffic data in bridge design. Washington: National Cooperative Highway Research Program, 2011 Contract No.: 683.

[79] Nowak AS, Hong Y-K. Bridge live-load models. Journal of Structural Engineering. 1991;117(9):2757–2767.

[80] O'Connor AJ, OBrien EJ. Traffic load modelling and factors influencing the accuracy of predicted extremes. Canadian Journal of Civil Engineering. 2005;32(1):270–278.

[81] Deng L, Yu Y, Zou Q, Cai CS. State-of-the-art review of dynamic impact factors on highway bridges. Journal of Bridge Engineering. 2015;20(5). DOI: 10.1061/(ASCE)BE.1943-5592.0000672

[82] OBrien EJ, Cantero D, Enright B, González A. Characteristic dynamic increment for extreme traffic loading events on short and medium span highway bridges. Engineering Structures. 2010;32:3827–3835. DOI: 10.1016/j.engstruct.2010.08.018

[83] OBrien EJ, Caprani CC, O'Connell GJ. Bridge assessment loading: A comparison of West and Central/East Europe. Bridge Structures. 2006;2(1):25–33.

[84] British Standard. Steel, concrete and composite bridges. Part 2: Specification for loads. London: British Standards Institution; 2006.

[85] The Highways Agency. Design manual for roads and bridges. Volume 3, Section 4, Part 3: The assessment of highway bridges and structures. London: The Stationery Office; 2001.

[86] Nowak AS. Calibration of LRFD bridge design code. Washington: National Cooperative Highway Research Program; 1999, Report No.: 368.

[87] Lipari A, OBrien EJ, Caprani CC. A comparative study of a bridge traffic load effect using micro-simulation and Eurocode load models. In: Biondini F, Frangopol DM, editors. 6th International Conference on Bridge Maintenance, Safety and Management. 2012; 8–12 July; Stresa. London: Taylor and Francis.

[88] Knospe W, Santen L, Schadschneider A, Schreckenberg M. Single-vehicle data of highway traffic: Microscopic description of traffic phases. Physical Review E. 2002;65(5).

[89] Dowling R, Skabardonis A, Carroll M, Wang Z. Methodology for measuring recurrent and nonrecurrent traffic congestion. Transportation Research Record: Journal of the Transportation Research Board. 2004;1867:60–68.

[90] The Council of the European Union. Official journal of the European Communities. Council Directive 96/53/EC of 25 July 1996 laying down for certain road vehicles circulating within the Community the maximum authorized dimensions in national and international traffic and the maximum authorized weights in international traffic 1996. Brussels. p. L235/59–75.

[91] The Council of the European Union. Official journal of the European Union. Directive (EU) 2015/719 of the European Parliament and of the Council of 29 April 2015 amending Council Directive 96/53/EC laying down for certain road vehicles circulating within the Community the maximum authorised dimensions in national and international traffic and the maximum authorised weights in international traffic 2015. Strasbourg. Strasbourg. p. L115/1–10.

[92] Treiber M, Kesting A. An open-source microscopic traffic simulator. IEEE Intelligent Transportation Systems Magazine. 2010;2(3):6–13.

[93] Committee on Loads and Forces on Bridges of the Committee on Bridges of the Structural Division. Recommended design loads for bridges. Journal of the Structural Division. 1981;107(7):1161–1213.

Chapter 3

3D and 4D Models Used in Bridge Design and Education

Alcínia Z. Sampaio

Additional information is available at the end of the chapter

http://dx.doi.org/10.5772/64675

Abstract

A bridge is the type of structure whose appearance normally deserves more attention because it not only has an evident impact on the environment but also represents considerable investment, both of which justify careful evaluation. The aesthetic analysis is an important issue that must be considered when designing a new bridge, especially when it is to be built in an urban or road environment. In this context, the automatic generation of three-dimensional (3D) geometric models of the bridge under analysis, and the walk around and aerial simulation allowed over it, which can be generated, helps bridge designers to evaluate its aesthetic concept and environmental impact. The bridge construction process can also be simulated, helping designers and builders to review the progress of the construction work in situ. For that, 4D (3D + time) models of the most frequent bridge construction methods were generated, using virtual reality (VR) technology. The simulation of the construction activity made possible by the developed interactive 4D model helps bridge designers to analyse the whole construction process. The present study aims to analyse the mechanisms of how to generate 3D models of a bridge automatically and how to simulate its construction using VR capacities.

Keywords: bridges, automation, 3D models, 4D models, construction, virtual reality

1. Introduction

The visual impact of an urban overpass or bridge inserted into a road network requires a careful study of its aesthetic aspect by its engineers. The importance of the aesthetic analysis of the infrastructure design in serving the public good demands special attention because of the densely built-up nature of the site where it is to be inserted that necessarily calls for order, discipline and aesthetic values [1]. Moreover, it is also important to analyse the environmental impact of a bridge, especially true of urban overpasses with regard to their more intrusive location close to areas of the pedestrian use.

Much of the public focus has been centred on several "landmark" bridges [2]. Denn [3] specifies aesthetic guidelines for bridge design, remarking that a careful and early application of aesthetic concepts can make a significant improvement in the appearance of bridges and structures. Every bridge is, to some degree, a historical document, a demonstration of structural technique, a performance test of building materials, a comment on the values of a society which produced it and a reflection of the richness or poverty of its designer's imagination. So, the bridge designer must strive to understand the creative process, together with scientific and technical principles. The aesthetic aspects that stimulate the senses in most viewers are proportion, order, simplicity, balance, colour and texture. Design bridges must incorporate these aesthetic principles with the physical and geometric components of the structure.

In this context, the generation of three-dimensional (3D) geometric models of bridges, which are to be designed and analysed, can play an important role. For this purpose, a computer graphic system, which enables the 3D geometric modelling of decks of the most frequent types of bridges, was developed. With this tool, the geometry of the bridge shape can be directly inserted into the computer application, using the user-friendly interfaces with geometric parameters of the longitudinal view and cross section of the bridge deck. In this way, the description of the geometry, conceived for each case, is easily achieved. In addition, it satisfactorily supports a rapid definition of several suitable alternative solutions for the bridge.

In addition to the 3D model of a bridge allowing its aesthetic analysis, it also supports the creation of 4D (3D + time) interactive models simulating the construction work. In order to create interactive 4D models, simulating the sequence and progress of the construction process, techniques of virtual reality (VR) were used. The designer links each construction task, established in a Gant map, to specific 3D model components, and programs the simulation of the bridge construction process, using the VR software. The virtual interactive 4D model allows users to view and interact with these construction stages and with the equipment involved in the process.

The incorporation of VR technology into 3D geometric models allows greater realism in the simulations; it is, now, often applied in the field of engineering perhaps because VR technology constitutes a good interface and provides the possibility of finding solutions to real-life problems in the construction field.

2. Automatic generation of 3D models

As aesthetic value is important for civil engineering projects, which impact urban and natural environments, it can be incorporated in curricular programs in engineering schools. However, the attention to structures traditionally dominates in the modern day university education of civil engineering, and the teaching of aesthetics meets obstacles. The main difficulty is the thought that aesthetic values are unfamiliar to engineers. This leads to the need for the formulation of main aesthetic thought into clear principles and rules that will help designers to learn and apply aesthetics to their design work.

Because aesthetics must be a part of the program for a bridge from the very beginning, this consideration must become the responsibility of every bridge engineer and must be an essential part of bridge engineering decisions at every step of the bridge design process; aesthetic consideration must be integrated carefully into the design of each detail of the bridge and its approaches [3].

To allow an easy and accessible evaluation of the visual appearance of a bridge, an intuitive tool to generate 3D models of bridges was developed. The application is an important tool to stimulate the diffusion of aesthetics in engineering projects and to simplify and support its analysis. The computational application allows the introduction of geometric data and organizes them in order to create three-dimensional models of the bridge. It also allows the user to generate the usual technical drawings required in the graphical documentation of a bridge design.

The computer application developed supports two bridge types: overpasses and box girder bridges. The computer application generates 3D models based on DXF format, an exchange code recognized by many CAD systems.

2.1. Overpass bridges

Overpasses are structural solutions that allow crossroads on different levels. They are usually used when a high-speed road or a road with intense traffic must be crossed, usually by a secondary road that passes above the dominant one (**Figure 1**).

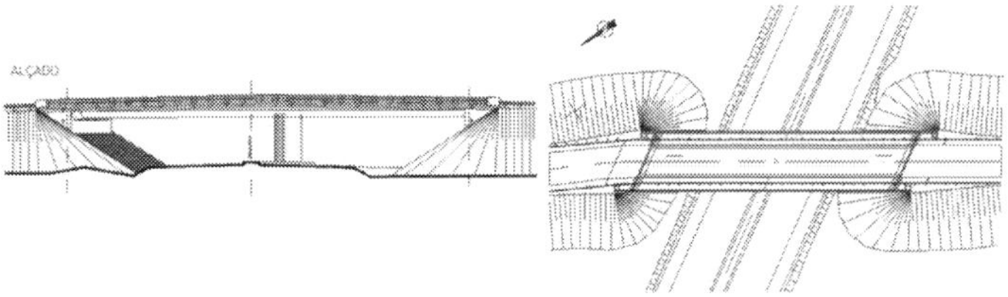

Figure 1. Overpass longitudinal and top views.

An overpass has a very significant visual impact but usually, such a careful aesthetic evaluation as that conducted for bridges is not considered necessary/worth the effort. The aesthetic analysis can be organized into two categories [3]: visual design elements and aesthetic design qualities. These structures have a high level of exposure, especially when they are in urban areas, so they should receive a special aesthetic analysis so as to provide a comfortable level of impact. The focus of the analysis is oriented to the shape of the bridge aesthetic aspects such as linearity, setting, brilliance, roughness and smoothness or aggression of shapes.

In the aesthetic study of an urban overpass, the implemented 3D modelling tool allows the easy creation of several solutions for the bridge, with distinct longitudinal shape and different type of deck cross section. The ability to create of 3D models easily, allowing the visualization of each alternative option, supports decision-making regarding the best solution for the bridge [4]. The interface of the overpass 3D modelling tool is composed of three main windows where the required parameters and supplementary information must be introduced. **Figure 2** shows the interface windows concerning the longitudinal deck shape and two types of cross-sectional shapes.

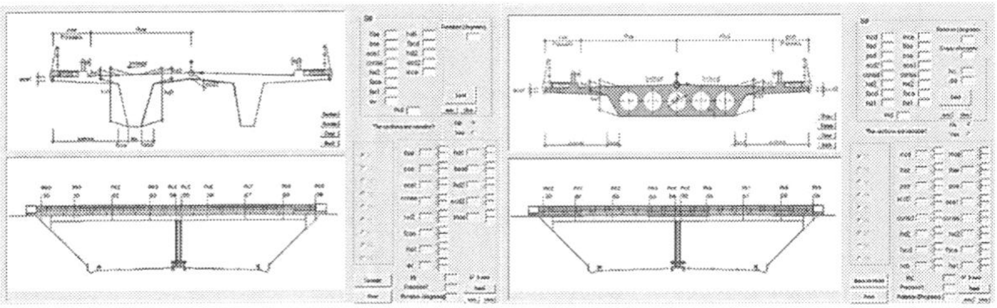

Figure 2. Interface of T-beam and slab section types.

The modelling process is based on geometric parameters allowing the definition of the bridge geometric configuration. It is not needed to indicate the vertices coordinates and the type of drawing element. Based on the parameter's values, the program allows the automatic generation of the desired bridge shape. So, the program makes possible to automatically generate 3D models. The parameterization was applied to two types of most used overpass sections, T-beam and slab (**Figure 2**); to linear and parabolic longitudinal variations of the deck and to the road geometry in plant and vertical views.

The program was created using Visual Basic programming language, and it is made up of three main windows. The cross-sectional window is presented in **Figure 2** and the horizontal and vertical alignment windows in **Figure 3**. The interfaces enable the definition of all the possible variations of the geometry of the road. For this purpose, the tool allows the geometric characterization of all design parameters needed, namely linear alignments, circular curves, transition curves and parabolic curves.

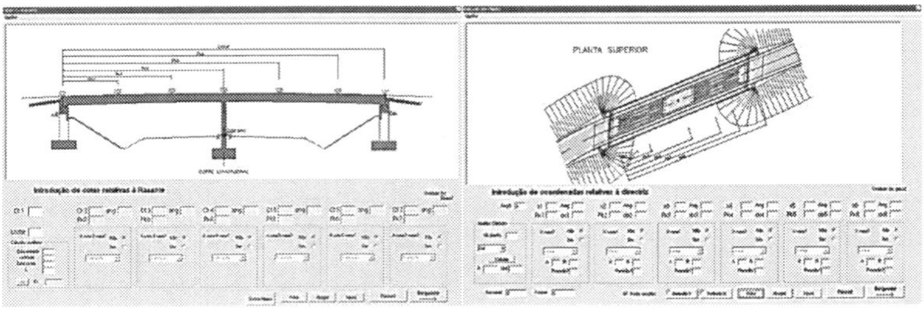

Figure 3. Vertical and horizontal alignment windows.

The program was tested in a case study. A first 3D model of the bridge was created, followed by several possible alternative solutions (**Figure 4**). The initial solution is a solid slab cross section with constant height, and the alternative options are as follows:

- a slab cross section with longitudinal linear height variation and parabolic height variation;
- a T-beam cross section with constant height, longitudinal linear height variation and longitudinal parabolic height variation.

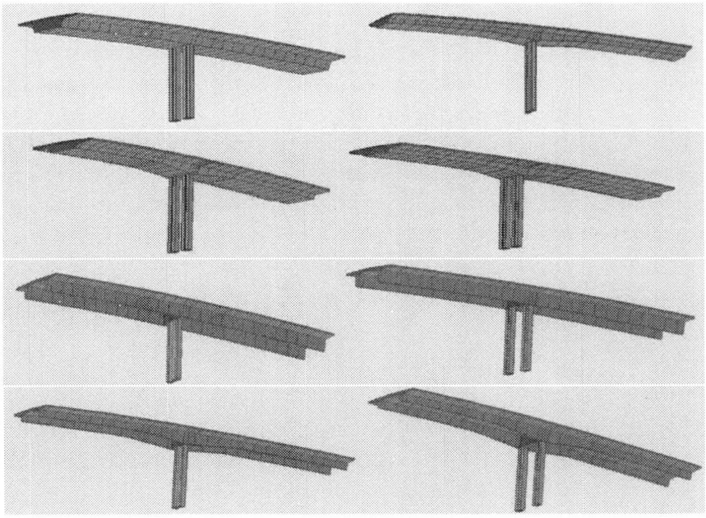

Figure 4. 3D model of the original and alternative solutions.

In addition, the program enables the consideration of as many pile forms as the engineer wishes for, with any longitudinal or transversal forms. This versatility made offered by the program facilitates a more thorough aesthetic study. The use of parametric computer programming makes it possible to quickly obtain the 3D model of the overpass and many alternative solutions under analysis. So, it allows users to carry out a more accurate study of the overpasses' aesthetics, supporting the comparative analyses between different possibilities and consequently the choice of the most suitable option. The modelling process should be done in steps, beginning with a global analysis of all the possible solutions, continuing with the consequent elimination of some of them and going on to more accurate and detailed analysis of a smaller group of possible solutions. This provides the opportunity to choose the solution that presents a good rhythm and a pleasant sense of continuity to the structure.

2.2. Box girder bridge

The same application has the capacity to generate 3D box girder deck superstructures [5]. The tool allows the creation of consecutive cross sections, exactly defined and correctly located, along the deck (**Figure 5**). The interface presents diagrams linked to dimensional parameters, in order to facilitate the description of the geometry established for each concrete case of the deck. The 3D model is the result of an appropriate combination of two longitudinal geometric components: the deck morphological evolution and the layout of the road, which acts simultaneously over a cross section, defining the exact deck shape'. The Annex item includes the algorithms used to calculate the 2D coordinates of each vertices of this type of cross section.

The geometric database, needed to create the 3D model of the deck, is formed by a set of geometric parameters, which the bridge designer deals with at the conceptual design stage. To obtain the 3D model of a deck segment, consecutive sections corresponding to the construction joints are generated using the algorithms of the program and connected to surfaces. The final configuration of the deck is comprised of two longitudinal surfaces: one representing the exterior side of the deck and the other its interior. The top cross sections are finally added to the 3D model (**Figure 5**). As this model uses cross sections correctly defined in shape and in their spatial orientation, it represents the real form of the deck. After the deck, 3D model is completed with the pillars and abutments, and it is inserted into the landscape. The aesthetic evaluation can then be carried out.

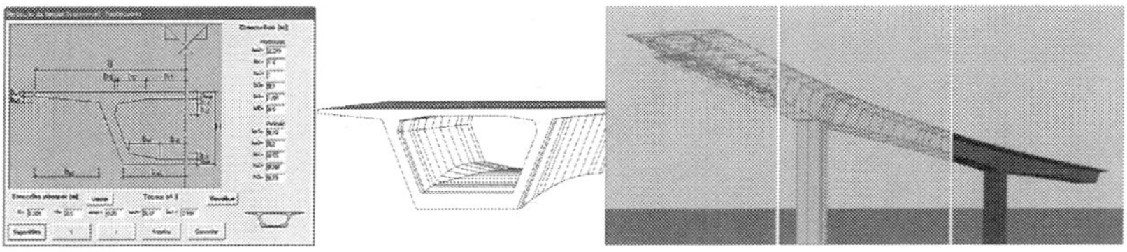

Figure 5. Box girder cross-sectional interface and 3D model.

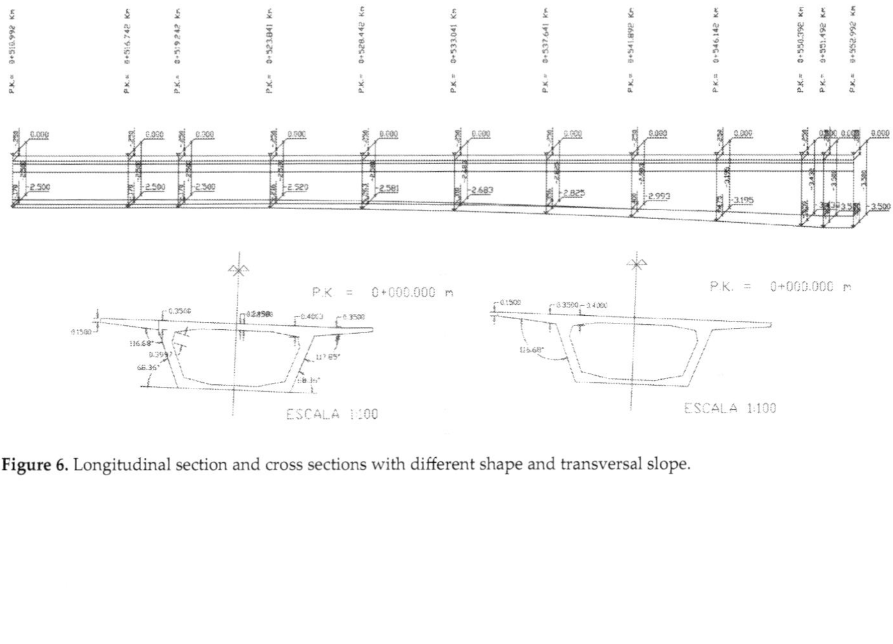

Figure 6. Longitudinal section and cross sections with different shape and transversal slope.

The implemented software allows the definition of traditional drawing usually required in a bridge project. As the horizontal alignment of the road is normally curved (**Figure 5**) and the road surface is never horizontal, the longitudinal deck section drawing must correspond in order to make a plan of the cut vertical surface defined along the deck axis (**Figure 6**). In this type of drawing, the elevation of the deck at distinct points along it needs to be calculated (**Figure 6**). These values depend on the elevation alignment characteristics. It is also a complex procedure to represent a set of consecutive cross sections with different interior and exterior shapes along the deck.

3. The simulation of the construction of a bridge

Technologies supporting 3D modelling and interaction with the models, which add to a better understanding in the teaching-learning in the classroom, have been introduced into schools [6, 7]. In particular, information and communication technology applications in higher education have been reported as improving learning, especially in course – based team learning and collaborative learning [8, 9]. In the field of education and training virtual reality (VR) technology has been used as a leading way for a better understanding of didactic issues, performed in face-to-face classes or in e-learning platforms. VR technology has proved very useful in the teaching of incremental processes [10]. The models that allow the visualization of the construction process of a building or bridge are 4D models, that is, they use the time factor linked to 3D construction components.

When implementing a 4D application, the designer must have a clear idea of what to show, because the objects to be displayed and the details of each one must be appropriate to the goal that the teacher wants to achieve with each specific model. In the development of the 4D bridge models, VR techniques were used, in order to improve their efficiency by allowing the interactivity by all parties involved in each type of bridge construction [11].

The 4D applications that were developed concern some of the most widely applied methodologies in bridge deck construction [12]. The 4D models allow users to demonstrate a process and present, briefly, the fundamental theory of the process or provide full information concerning the experiments. In the field of civil engineering, there are several construction methods for the building of bridges. The most frequent constructive processes for bridge decks are (**Figure 7**): bridge deck formed by precast beams [13]; cantilever construction [12] and the incremental launching method [14].

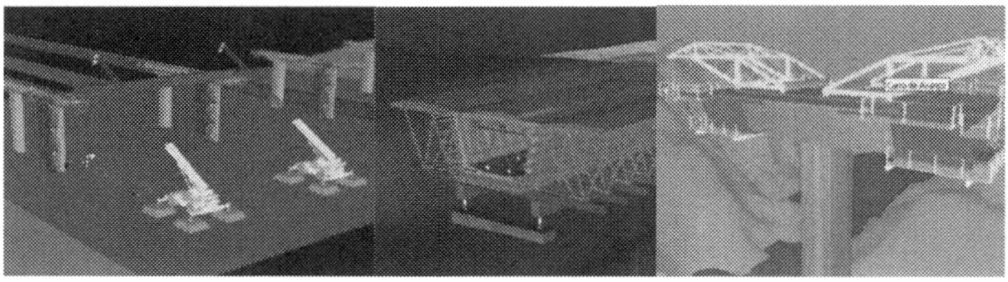

Figure 7. Methodologies of deck bridge construction.

In order to establish the principal characteristics of each construction method, to define the type and quantity of the construction components needed to be modelled for the simulation of each construction work and, also, to perform the correct sequence of operations, the bibliographies concerning the three methodologies of bridge construction were consulted [12–14]. Based on this information, the 3D model of each type of bridge was generated and the corresponding sequence animation of each construction process was programmed. Because specialists in construction processes and bridge design were consulted in the implementation of these 4D models, the final product is efficient and accurate. Through direct interaction with the models, the progress of the actual construction process of the bridges can be monitored.

3.1. Bridge deck composed of prefabricated beams

The construction method for decks composed of prefabricated elements is frequently chosen in construction works over railway lines, and, in general, in areas where the placement of trusses is difficult. This is because these bridges offer several advantages in urban areas, as they allow quick and economical construction without generating significant local constraints [1]. In a bridge deck, the prefabricated beams present an equidistant distribution. The slab complemented "in situ" uses pre-slabs as lost shuttering and as reinforcement during service, contributing to the structural strength of the deck. The most common cross-sectional type of the prefabricated beams is an I-shape.

The first step of the constructive method consists of placing the prefabricated beams on the pillars and can be carried out by means of cranes (**Figure 8**). The connection between the beams is made using pre-slabs.

The 4D virtual model that was implemented presents the construction of a deck composed of 4 precast I-beams, lifted by cranes onto the pillars and supplemented with composite pre-slabs [13]. Initially, 3D geometric models of all the elements necessary to simulate the construction process were created, including the surroundings of the construction site, the pillars, the stair towers, the worker platforms, the provisional and the definitive supports and two cranes needed to lift the precast beams (**Figure 9**). The 3D model of a prefabricated beam includes reinforcements running out of the beam.

Figure 8. Placement of precast beams and pre-slabs.

http://dx.doi.org/10.5772/64675

Figure 9. 3D models of beams, pillars, stair towers and worker platforms.

The virtual simulation of the construction activity starts with the presentation of the workplace, followed by the insertion of additional elements, such as the stair towers (for access to the top of the pillars) and the work platforms (which allow the workers to move around and complete their tasks). The sequences of activities in the virtual space are as follows (**Figure 10**):

- the definitive and temporary supports are placed on the top of the pillars;
- the beams are raised by two cranes and placed on the support elements;
- next the pre-slabs are placed on the prefabricated beams;
- the reinforcements of the slab are placed, and finally, the deck slab and the transversal beams are concreted.

Figure 10. Construction sequence deck composed with prefabricated beams.

After finalizing the construction of the deck, the provisional support devices are removed, and all complementary elements necessary for road traffic are placed above the deck. The complete bridge can now be observed from any point of view (**Figure 11**). The model allows the user to use the zoom sufficiently well in order to understand the final configuration of the bridge.

Figure 11. Views of the complete deck.

3.2. Incremental launching method

Another interactive model for the construction of bridge decks was created. The construction of bridge decks using the incremental launching method has existed from the 60s. This method consists of casting segments of the bridge deck in a provisional formwork, and then, each segment is pushed forward along the bridge axis. This method is used in viaducts crossing high valleys. The cross section of the bridge must have a constant height, and the most suitable type of cross section is the box girder.

Using the VR model is possible to follow the visual simulation of the construction sequence and to learn how the equipment is moved [15]. In order to perform correctly, the construction simulation activity every construction components and equipment was generated as 3D models (**Figure 12**), and the EON studio, a VR-based software, was used [16].

The main steps of the construction process are as follows (**Figure 13**):

- the external panels of the shuttering are first placed followed by the reinforcement mesh;
- after, the interior false work is placed incrementally in sequence, starting with the metallic element, followed by the longitudinal beams, the shuttering panels and finally the launching nose are assembled;
- a first segment is casted, the nose is removed, the segment is separated from the shuttering, and displacement of the element takes place.

For the following segments, the process is identical. In the final phase of the construction the worksite yard is removed. Finally, all the finishing elements are positioned. Because consideration was given during its development, both to technical knowledge and to its use in education, in particular how and what to show, the 4D model could be an important teaching tool to illustrate bridge construction issues during training. The application is designed, not only as a learning tool, but also for use by professionals involved in the construction of these types of bridges. Note that a film was created showing the interaction with the VR/4D model (available in [14]).

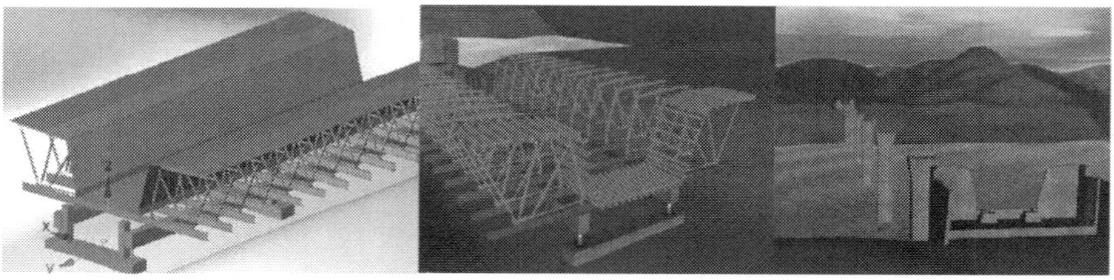

Figure 12. 3D models of the construction elements and landscape.

Figure 13. Sequence of the incremental launching process.

3.3. Cantilever bridge deck construction

A further 4D model allowing the simulation of the cantilever method was created [17]. Students and teachers are able to dictate the speed of the process, in order to observe the movement of advanced equipment and how to place each elements of the bridge. The sequence performance is programmed using the VR software followed by an appropriate planning guides for this type of construction.

The case study has a box girder cross section, and its height varies parabolically along three spans. The most common construction technique for this typology is the cantilever method of bridge deck construction.

The aforementioned computer graphic system allows the generation of the deck segments needed for the construction simulation of the bridge [5]. To complete the virtual scenario of the construction site, the advanced equipment, the formwork adaptable to the size of each segment, the platforms for the workers and the false work to be placed near the abutments were also modelled (**Figure 14**).

Figure 14. 3D models of a deck segment, the scaffolding and the advanced equipment.

The input and support of bridge designers, not only on the geometric definition of the bridge components and devices, but also on the establishment of the progression sequence and the way the equipment operates (**Figure 15**) were essential to obtain an accurate model:

- a first segment is concreted on each pillar and is installed the work equipment on it;
- the process of concreting segments is defined in symmetrical way, starting from each pillar and using the advanced equipment;
- the continuity of the deck is established with a closing segment, using just one advanced equipment;
- finally, near the abutments, the deck is constructed, using a false work.

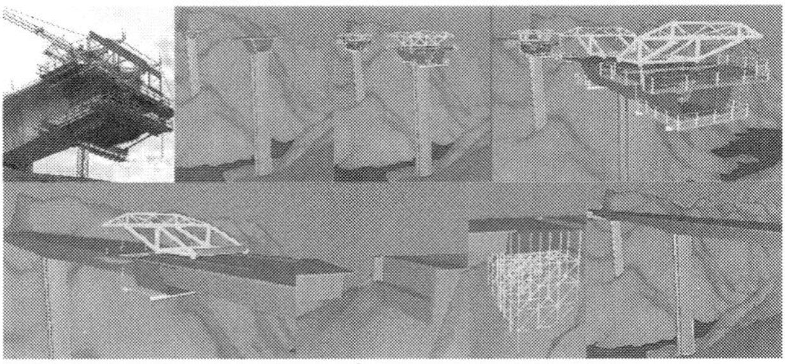

Figure 15. Sequence of the cantilever construction process.

In a real construction site of a bridge, for security reasons, the student is obliged to stay far from the zone where a bridge is under construction and thus cannot observe, in detail, the methodologies or the progression of the construction. However, while using the interactive virtual model, by moving the camera closer to the virtual bridge and applying to it routes around the zones of interest, the student can follow the sequence specifications and observe the details of the configurations of the construction components. Being able to interact with the bridge models therefore should help students gain better understanding.

3.4. Integration of interactive capacities

The attribution of virtual properties to the models in each application was defined by using the VR-based tool, the EON studio [16]. The implementation of the interactive model, con-

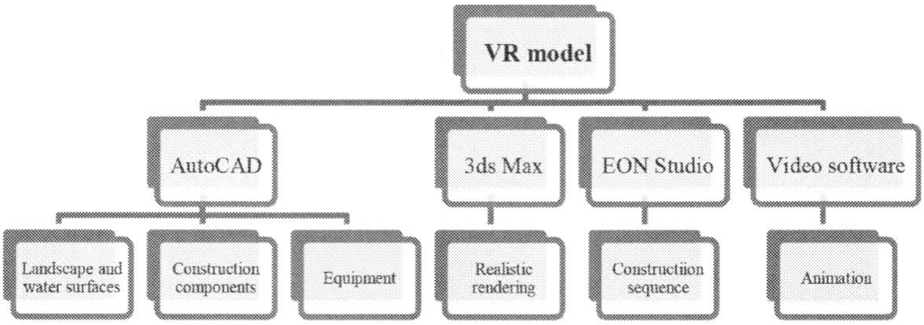

Figure 16. Software used on the implementation of a VR model.

cerned with the incremental launching method, is based on several linked applications as shown in the chart in **Figure 16**.

In the implementation of the 4D model for cantilever method, 3D models of the box girder bridge and of the environment were transposed (as 3D Studio Objects in **Figure 17**) to the EON Studio. The EON system is object programming-based software: the nodes window contains all actions that can be associated to the elements included in the simulation tree window. For instance, the advanced equipment model is identified, in the simulation tree shown in **Figure 17**, as a group or frame, designated as a "car."

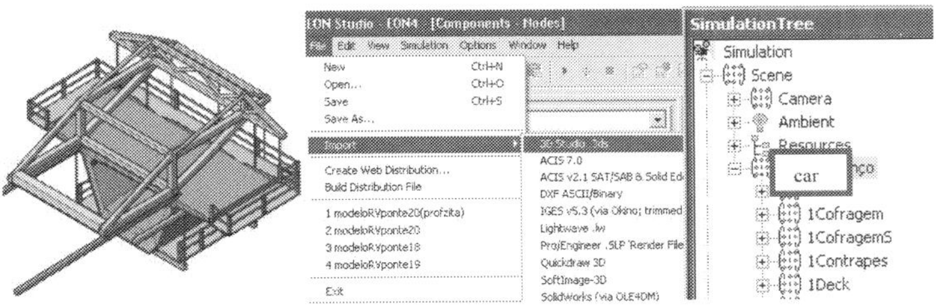

Figure 17. 3D model of the advanced equipment in 3DS format and the EON system main interface.

The definition of the construction sequence is based on a *counter* option button, which determines the next action in the process. The first action consists of the insertion of pillars in the landscape. When the components of the model are transposed to the EON, every element, except the landscape, is associated with the hidden characteristic The order to display an element is commanded by the counter node (**Figure 18**), which contains the logical instruction, which means when the value in the *counter* is accurate, then the associated element will be displayed. Consequently, when the mouse is clicked, the counter indicates the next element to be displayed. The correct programming will simulate the steps of the real bridge deck construction.

Figure 18. Counter node.

3.5. Educational considerations

The aim of the practical application of the VR models is to provide support in civil engineering education, particularly in those disciplines relating to bridges and construction processes in classroom-based education. They can also be used to assist professional training and distance learning based on e-learning technology. When this visualization tool was being designed, human perceptual and cognitive capabilities were taken into account [18]. It means that the program is suitable for use in a wide range of learning environments or stages of education.

The traditional way to present the curricular subjects involved in these virtual models is through 2D layouts or pictures. By using the 4D models, teachers may help the students to visualize and engage with the construction process more interactively. The following are some advantages of using the 4D models:

- The 4D bridge models show the complexity associated with the construction work on the deck. The models also illustrate, in detail, the movement of the equipment. In class, the teacher can/has to explain how the process must progress and the way the equipment operates (**Figure 19**).
- The incremental method model presents the geometry of the different elements and the materials used in real work processes in all their complexity. The camera movement shows the model in a consistent manner when presenting the sequences of all events that are programmed. It allows the user to acquire a correct understanding of the most important aspects of the construction method. In addition, to allow consultation of the required data in any phase, information about the construction process is integrated into each new component or step (**Figure 20**).
- The fact that the models are going to be used by undergraduate civil engineering students was taken into consideration. The models support the learning process by illustrating the bridge construction issues and by presenting the geometric details involved and the technical information that must go with each constructive step. The details available in the 4D models complement the educational applications, rendering them more useful and efficient. The students will get better understanding of the construction operation and be able to make full use of this knowledge when they go on an educational visit to the construction site as they will have been previously exposed to the essential information This type of model allows the participant to interact in an intuitive manner with the 3D space, to repeat the sequence or task until the desired level of proficiency or skill has been achieved, with the extra advantage of this practice being carried out in a safe environment.

The introduction of the VR model as a new teaching tool in construction and bridge disciplines has been well accepted, although some difficulty in the manipulation of the model was reported. However, this kind of new technological material, based on 3D/4D interactive models, is important in a modern class setting and deserves attention.

Teachers are the key players in the educational process and are the main determiners of quality in the classroom, so they must be kept up to date with new technological material that can contribute to the enhancement of quality in education. Therefore, this new concept of VR technology applied to educational models can bring new perspectives to the teaching and learning process for civil engineering education.

Figure 19. The models show in detail the movement of the equipment during the construction process.

Figure 20. The menu of events supports interactivity with the model.

4. Conclusions

Aesthetics has an important role to play in civil engineering education and must become an integral part of all civil engineering disciplines in order to minimize their possible negative impact on the environment and maximize the possibilities of improving it. The world is becoming more sensitive to this necessity, and the range of tools at our disposal for this is increasing. The new technologies are becoming more powerful, and we should make them more usable. This work describes the implementation of software that contributes to the visual analysis of bridge structures. A generative program to define bridge decks with several cross-sectional types was created by using 3D geometric bridge deck models. The usefulness and applicability of the program were successfully tested on several cases showing that the 3D models are simple to create and to adapt with those modifications clearly needed during the development of the bridge design.

In addition, the text describes the implementation of interactive models that simulates the construction work activity of some constructive processes concerning bridge deck methodologies. Virtual reality (VR) capacities applied over the 3D models of the bridges allow interaction with the construction process contributing to better understanding of the whole construction methodology. The model 4D (3D + time) offers important advantages in the construction field, allowing the student to learn about the composition of a bridge, the sequence established for each construction method and observe the movement of the equipment.

The text shows the application of new technologies to stimulate the diffusion of aesthetics in engineering projects, to follow the construction process and to simplify bridge analysis. The bridge VR/4D tools developed can be considered as improvements when compared with traditional ways of presenting the bridge construction issues in support of the construction activity and of training in these processes.

The main objective of this practical application of the model is to support the understanding of the construction process and to assist training professionals. This chapter shows that the applications with these characteristics make the advantage of using techniques of virtual reality more self-evident, especially when compared to the simple manipulation of complex models, which cannot be broken down, in it a useful contribution to the field.

Annex

The parametric cross sections of the box girder deck and algorithms to obtain the (x, z) coordinates of each vertices.

The geometric parameters set is {*Be, be, h, ve1, ve2, ve3, ve4, ve5, he1, he2, he3, he4*}. The coordinates x and z of each vertices of the box girder cross section are obtained using the following algorithms.

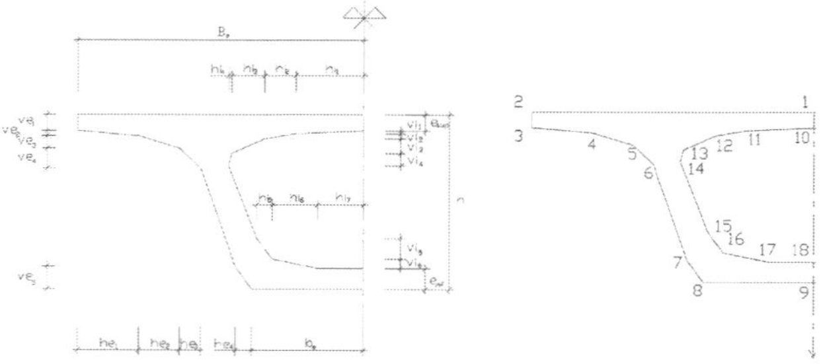

x(1) = 0	z(1) = 0	x(2) = -Be	z(2) = 0
x(3) = -Be	z(3) = ve1	x(4) = x(3) + he1	z(4) = z(3) + ve2
x(5) = x(4) + he2	z(5) = z(4) + ve3	x(6) = x(5) + he3	z(6) = z(5) + ve4
x(7) = -be - he4	z(7) = h - ve5	x(8) = -be	z(8) = h
x(9) = 0	z(9) = h	x(10) = 0	z(10) = esup
x(11) = x(10) - hi1	z(11) = z(10) + vi1	x(12) = x(11) - hi2	z(12) = z(11) + vi2
x(13) = x(12) - hi3	z(13) = z(12) + vi3	x(14) = x(13) - hi4	z(14) = z(13) + vi4
x(18) = 0	z(18) = h - einf	x(17) = x(18) - hi7	z(17) = z(18)
x(16) = x(17) - hi6	z(16) = z(17) - vi6	x(15) = x(16) - hi5	z(15) = z(16) - vi5

Author details

Alcínia Z. Sampaio

Address all correspondence to: zita@civil.ist.utl.pt

Department of Civil Engineering, University of Lisbon, Lisbon, Portugal

References

[1] Podolny W Jr. An overview of precast prestressed segmental bridges. PCI Journal. 1979. Available from: https://www.pci.org/Design_Resources/Guides_and_Manuals/References/Bridge_Design_Manual/JL-79-January-February_An_Overview_of_Precast_Prestressed_Segmental_Bridges/, [Accessed 10th January, 2016].

[2] Leonhardt F. Aesthetics of bridge design Technische Hochschule Stuttgart, Stuttgart, Germany. PCI Journal. 1978. Available from: https://www.pci.org/design_resources/guides_and_manuals/references/bridge_design_manual/jl-68-february_aesthetics_of_bridge_design/ [Accessed 27th July, 2016].

[3] Minnesota Department of Transportation, USA. Aesthetic guidelines for bridge design. 1995. Available from: http://www.dot.state.mn.us/bridge/pdf/aestheticguidelinesforbridgedesign.pdf [Accessed 10th January, 2016].

[4] Costa AA. Aesthetic analyses of viaducts supported on the automatic generation of 3D models [thesis]. Lisbon: University of Lisbon. 2007. 80 p.

[5] Alcinia Z. Sampaio Geometric modelling of box girder deck for integrated bridge graphical system. Automation in Construction. 2003; 21:55–66. PII: S0926-5806(02)00040-7.

[6] Goebbels G, Göbel M, Hinkenjann A, Heiden W, Winzker M, Herpers R, Scholl R, Reinert D. Institute of Visual Computing – Bonn–Rhine–Sieg University of Applied Sciences, Germany. SBC Journal on 3D Interactive Systems. 2011; 2(2). Available from: http://vc.inf.h-brs.de [Accessed 10th January, 2016].

[7] Crosby-Nagy M, Carfora J. ICT applications in US higher education. In: Vincenti G, Braman J, editors. Teaching through Multi-User Virtual Environments: Applying Dynamic Elements to the Modern Classroom. 2011, Chapter 4. pp. 47–58. Available from: http://www.igi-global.com: Information Science Reference (an imprint of IGIGlobal), ISBN: 978-1-61692-822-3, [Accessed 10th January, 2016].

[8] Haque ME, Suketu Shah PE, Agarwal G. A virtual tour of energy conscious architecture. In: 2005 American Society for Engineering Education Annual

Conference and Exposition. 2015. Available from: http://architectural.asee.org/2005_papers/2005-938_Final.pdf, [Accessed 27th July, 2016].

[9] Berka R, Bittner J, Slavık P. SBC Journal on 3D Interactive Systems, Special Issue—VR and HCI Labs. VR and HCI Labs at the Czech Technical University in Prague [Internet]. 2011. Available from: http://seer.ufrgs.br/jis [Accessed 10th January, 2016].

[10] Irizarry J, Meadati P. Use of interactive display technology for construction education applications. In: 2009 ASEE Southeast Section Conference. 2009. Available from: http://155.225.14.146/asee-se/proceedings/ASEE2009, [Accessed 21st July, 2016].

[11] Marini D, Folgieri R, Gadia D, Rizzi A. Virtual reality as a communication process. In: Virtual Reality, Editors: D. Ballin; R.D. Macredie. London: Springer; 2011. p. 1359–4338. doi:10.1007/s10055-011-0200-3. Available from: http://link.springer.com/article/10.1007/s10055-011-0200-3.

[12] Sampaio AZ, Ferreira MM, Rosário DP, Martins OP. 3D and VR models in civil engineering education: construction, rehabilitation and maintenance. Automation in Construction. 2010; 197: 819–828. doi:10.1016/j.autcon.2010.05.00

[13] Viana LF. Simulation of the Construction of the Bridge Deck Composed of Prefabricated Beams based on Virtual Reality Technology [thesis]. Lisbon, Portugal: University of Lisbon. 2012. 80 p.

[14] Octavio Martins. Incremental launching of bridges [Internet]. March 2015. Available from: http://www.octaviomartins.com/lancamentoIncremental/ [Accessed 10th January, 2016].

[15] Martins OP, Sampaio AZ. Virtual visual simulation of the incremental laughing method of bridges construction. In: 17° Portuguese Meeting of Computer Graphics. Covilhã, Portugal: University of the Interior Beira. 2009. pp. 29–33.

[16] EON Reality, Inc., Introduction to working in EON Studio [Internet]. 2012. Available from: http://www.eonreality.com/ [Accessed 10th January, 2016].

[17] Sampaio AZ, Studer P. Learning construction processes using virtual reality models. ITcon Journal—Electronic Journal of Information Technology in Construction. 2005; 10:141–151, paper 11, ISSN: 1400-6529. Available from: http://www.itcon.org/2005/11, [Accessed 27th July, 2016].

[18] Haquen ME. nD virtual environment in construction education. In: The 2nd International Conference on Virtual Learning, ICVL 2007; October 26–28, 2007. Constanta, Romania: University OVIDIUS Constanta. pp. 81–87. Available from: http://pt.scribd.com/doc/14068823/Proceedings-of-ICVL-2007, [Accessed 27th July, 2016].

Chapter 4

The Structural Performance of Stone-Masonry Bridges

George C. Manos, Nick Simos and
Evaggelos Kozikopoulos

Additional information is available at the end of the chapter

http://dx.doi.org/10.5772/64752

Abstract

The structural performance of old stone-masonry bridges is examined by studying such structures located at the North-West of Greece, declared cultural heritage structures. A discussion of their structural system is included, which is linked with specific construction details. The dynamic characteristics of four stone bridges, obtained by temporary *in situ* instrumentation, are presented together with the mechanical properties of their masonry constituents. The basic assumptions of relatively simple three-dimensional (3-D) numerical simulations of the dynamic response of such old stone bridges are discussed based on all selected information. The results of these numerical simulations are presented and compared with the measured response obtained from the *in situ* experimental campaigns. The seismic response of one such bridge is studied subsequently in some detail as predicted from the linear numerical simulations under combined dead load and seismic action. The performance of the same bridge is also examined applying 3-D non-linear numerical simulations with the results used to discuss the structural performance of stone-masonry bridges that either collapsed or may be vulnerable to future structural failure. Issues that influence the structural integrity of such bridges are discussed combined with the results of the numerical and *in situ* investigation. Finally, a brief discussion of maintenance issues is also presented.

Keywords: stone-masonry bridges, structural performance, *in situ* measurements, numerical simulations

1. Introduction

This chapter focuses on stone-masonry bridges that were built in Greece during the last 300 years and most of them survive today (**Figure 1a** and **b**).

Figure 1. (a) Konitsa Bridge, Ipiros, Greece and (b) Kokorou Bridge, Ipiros, Greece.

Figure 2. The royal tomb of Atreus at Mycenae in Peloponnese, Southern Greece.

The use of the stone-masonry arch that is utilized in forming stone-masonry bridges was extensively used in the times of the Roman Empire as part of the transportation system that was established and linked to various provinces of the Roman Empire. Evidence of stone-masonry arch bridges prior to Roman times is not known although stone-masonry structures in the East Mediterranean area for other uses date to prehistoric times. A well-known use of arch/vault stone-masonry structural form is the one that can be seen at the royal tombs, which have been excavated during the last 200 years in many places in Greece. In **Figure 2**, the royal tomb of Atreus at Mycenae, Greece, is depicted where stone masonry is employed to form an underground-vaulted structure with a diameter at its base of 14.60 m and a height of 13.30 m constructed with 33 subsequent series of stone masonry along the height.

The use of such vaulted stone-masonry structures demonstrates the efficient utilization of this structural form in order to bear efficiently the dead loads as well as the weight of the overlying soil volume in a state of stress dominated by compression (**Figures 2** and **3a**). On the contrary, the main gate of the royal palace walls at Mycenae in Peloponnese of Southern Greece (dated 1325 B.C. to 1200 B.C. and excavated 150 years ago), known as the gate of the lions (**Figure 3b**), uses the simple-supported beam-type structural system that characterizes most of the prehistoric and classical ancient Greek stone-masonry construction for above-the-ground structures. The use of the stone-masonry arch/vault-type formation is also evident in the structural system of the royal tombs of the Macedonian kings at Vergina in Northern Greece, dated from 350 B.C. and excavated during the last 30 years (**Figure 3c** and **d**).

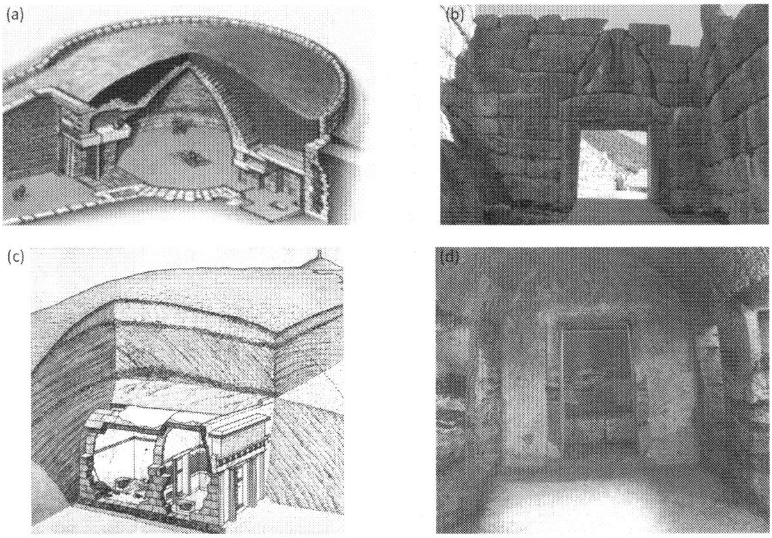

Figure 3. (a) Reconstruction of the royal tomb of Atreus, at Mycenae in Peloponnese, Southern Greece. (b) The gate of lions at Mycenae, Peloponnese, Southern Greece. (c) Reconstruction of the royal tomb of Philip, King of Macedonia at Vergina, Greece. (d) Interior of a Macedonian, royal tomb, Greece.

Figure 4. (a) Map of Macedonia with the location of the royal palaces at Vergina and Pella, Greece. (b) The remains of an ancient Roman bridge at a distance of 25 km from the Macedonian palaces of Vergina and Pella.

Despite the use of arch/vaulted stone-masonry structural formations for these underground Macedonian royal tombs at Vergina in Northern Greece, there is no evidence of such structural formations being used for bridges at that time. **Figure 4a** shows the location of the Macedonian royal palaces at Vergina and Pella in Northern Greece (red arrows).

In the same figure, the location of the remains of an ancient Roman bridge (blue arrow) is also indicated. These remains correspond today to only one main arch with a span of 15 m and a height of 7.5 m (**Figure 4b**). This surviving part of a Roman stone-masonry bridge is dated between 50 A.D. and 150 A.D. and, as can be seen in the map of **Figure 4a**, is located at a close distance (25 km) from the Macedonian palaces of Vergina and Pella as well as for the important cities of Thessaloniki and Dion (30–40 km). An inventory of Roman stone-masonry bridges is given by O'Connor [1]. These structures survive today, located in many European countries, having been in many cases preserved in good condition (**Figure 5a** and **b**) or partially collapsed in other cases (**Figure 5c** and **d**).

Figure 5. (a) Roman stone-masonry bridge Pont-Saint-Martin in Northern Italy. (b) The Pont Julien, a Roman stone arch bridge in the southeast of France, dating from 3 B.C. (c) The Pont Ambroix, first century B.C., Roman bridge in the south of France damaged by severe floods. (d) Ponte Rotto/Emilio, Rome (Broken/Emilio bridge). The remains of stone-masonry bridge damaged by flooding.

2. Geometric characteristics of the stone-masonry bridges located at North-West Greece

In what follows, a brief review is given of the basic geometric and construction characteristics of the stone-masonry bridges located at the far North-Western part of Greece called Ipiros. Bridges of similar geometric and construction characteristics are also located in other parts of Greece. The present study has selected the stone-masonry bridges that are located in Ipiros as they are numerous and are located in a relatively confined area that facilitates their temporary *in situ* instrumentation. The objective of this instrumentation, as explained in Section 4, is to measure their dynamic characteristics that represent a significant part of this study. All these

http://dx.doi.org/10.5772/64752

bridges, located in Ipiros as well as in other parts of Greece, have been documented, with relevant information included in [2]. Psimarni et al. [3] developed a geographic information system for the traditional bridges of Central Zagori, not yet accessible to the authors. Thus, all the geometric data utilized in this study were obtained through *in situ* measurements conducted by the authors.

Figure 6. (a) Kapetan Arkouda Bridge, Kipoi Village, East Zagori, Ipiros and (b) Agiou Mina Bridge, Kipoi Village, East Zagori, Ipiros.

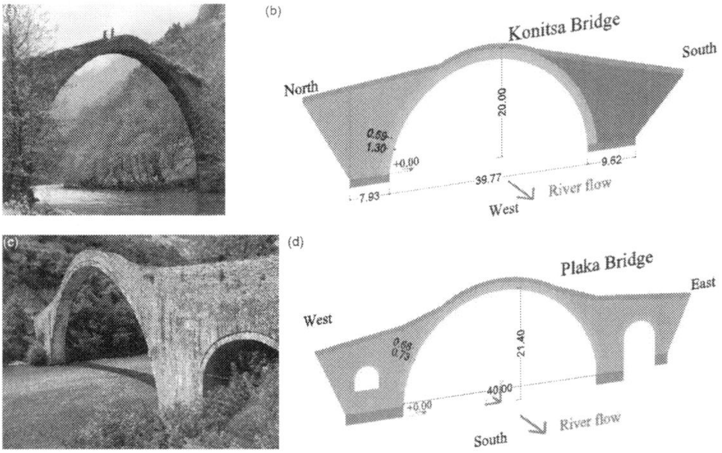

Figure 7. (a) and (b) Konitsa Bridge, Ipiros, Greece (width 2.85 m); (c) and (d) Plaka Bridge, Ipiros, Greece (width 3.10 m).

A considerable number of relatively small stone-masonry bridges can be found in this region with a span smaller than 10 m as the ones depicted in **Figure 6a** and **b**. However, stone-masonry bridges with a much larger total span have also been constructed. Relatively long-span stone bridges with a single central span are relatively few in number. The longest stone bridges with one main central arch are the ones in Konitsa (**Figure 7a** and **b**) and the one in Plaka (**Figure 7c** and **d**). These stone-masonry bridges are very similar in the dimensions of the central arch, although the Arachthos river crossing by the Plaka Bridge is longer (75-m total span) due to adjacent additional arches at both ends (**Figure 7c**), whereas the main arch of the Konitsa Bridge is supported directly at the nearby slopes of the rocky Aoos river gorge. As will be presented briefly in Chapter 9 (see also figures 10g and 11 as well as section 7.3), the Plaka Bridge collapsed almost a year ago (31 January 2015). As can be seen in **Figure 7b** and **d**, the main central arch of both the Konitsa and the Plaka stone bridges has a clear span of nearly 40 m and a rise of 20 m.

Apart from the Plaka and Konitsa stone bridges, three more bridges will be examined in the present study. These stone bridges are depicted in **Figure 8a–f** and are namely the Kokorou Bridge, the Tsipianis Bridge and the Kontodimou Bridge. As can be seen in **Figure 8b** and **d**, the main central arch of the Kokorou Bridge has a clear span of 24.69 m and a rise of 12.71 m, whereas the Tsipianis Bridge has a clear span of 26.00 m and the rise 13.65 m. As can be seen, the central main arch of these two bridges has similar dimensions. Finally, the clear span of the Kontodimou Bridge is 14.50 m and the rise 7.40 m.

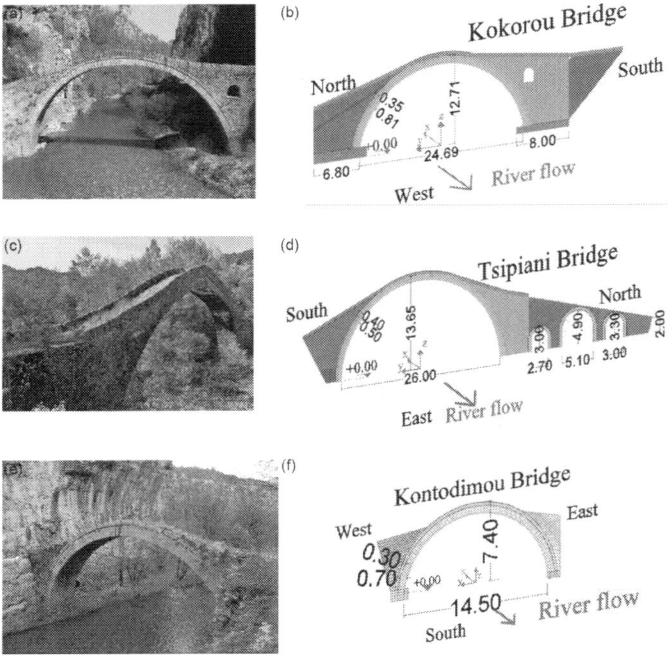

Figure 8. (a) and (b) Kokorou stone bridge, Kipoi Village, East Zagori, Ipiros (width 2.85 m); (c) and (d) Tsipianis stone bridge, Milotades Village, East Zagori, Ipiros (width 2.80 m); (e) and (f) Kontodimou stone bridge, Kipoi Village, East Zagori, Ipiros (width 2.77 m).

http://dx.doi.org/10.5772/64752

Thus, the present study covers stone bridges that are all dominated by a central main arch with a span/rise varying from 40.00/20.00 to 14.50/7.40 m. As can be seen in all cases, the clear span over rise ratio is close to 2.0 and the width is close to 3.0 m. In the cases of the longest span, the width of the structure increases as the arch approaches the foundation (Konitsa and Plaka Bridges). A distinct difference between the examined bridges is the fact that in the case of Konitsa, Kokorou and Kontodimou Bridges, the main arch is founded on abutments that are very close to the rocky slopes of the river gorge whereas for the Plaka and Tsipianis Bridges there is a mid-pier that is founded on the river bed together with adjacent smaller arches (**Figures 7d** and **8d**).

3. Construction characteristics

The construction characteristics of the various parts of these bridges are thought to bear some significance in the effort to understand the static, dynamic and earthquake behaviour of these structures. One can distinguish the following main structural components:

Figure 9. (a) Main central arch supported on the extension of the rocky part of the river bank at both ends; (b) main arch supported on right and left mid-piers which are also formed including adjacent arch; (c) wooden formwork for the support of the main central arch. Construction of the right and left abutments; and (d) construction of the main central arch preceded by the construction of the right and left abutments.

1. The main primary central arch is founded on abutments at either end of the bridge. These abutments are extensions of the rocky part of the river bank (**Figure 9a**). In the case of a

mid-pier, which was constructed on the dry part of the river bed, a separate foundation footing is constructed at a certain depth that is not easy to estimate. In case of relatively large river widths, the main arch was founded on right and left mid-piers that were also supporting adjacent arches as is seen in **Figure 9b**. A wooden formwork was employed to support the main arch during construction (**Figure 9c**).

2. The construction of the main central arch was preceded by the construction of its foundation at both ends together with the construction of the abutments that were raised up to a certain height in order to resist the thrust of the central arch.

3. The construction of the main central arch was followed in many cases with the construction of a secondary central arch on top of the main central arch (**Figures 9d** and **10a–f**).

4. Finally, the mandrel walls were constructed above the abutments in order to form together with the arches the main passage (deck) at the top of the bridge. In certain cases, this passage is protected at both sides at the deck level by an in-built continuous stone parapet that rises approximately 0.5 m above the deck level (**Figures 7a, c** and **8a, c**). In the case of the Kontodimou Bridge, this parapet is formed by individual stones in-built at intervals of approximately 1.6 m.

5. The thickness of the primary and secondary arches of the main span varies considerably. The primary arch for the Konitsa Bridge with a clear span of 40 m has a thickness of 1.30 m And The Secondary Arch A Thickness of 0.59 m (**Figures 7b** and **10a**). The primary arch of the Plaka Bridge again with a clear span of 40 m has a thickness of 0.73 m and the secondary arch a thickness of 0.68 m (**Figures 7d** and **10b**). The primary arch of the Kokorou Bridge with a clear span of 24.69 m has a thickness of 0.81 m and the secondary arch a thickness of 0.35 m (**Figures 8b, 10c** and **d**). The primary arch of the Tsipiani Bridge with a clear span of 26 m has a thickness of 0.50 m and the secondary arch a thickness of 0.40 m (**Figure 8d**). Finally, the primary arch of the Kontodimou Bridge with a clear span of 14.5 m has a thickness of 0.70 m and the secondary arch a thickness of 0.30 m (**Figures 8f, 10e** and **f**). These thickness values are approximate and correspond to the arch thickness at the maximum rise; in some cases, the primary and the secondary arch thicknesses vary having an increased thickness in the areas where these arches join the abutments.

6. The construction of both the primary and secondary main central arches as well as the rest of the arches was constructed with stones that were shaped in a very regular prismatic shape. In this way, the mortar joints of the masonry construction for these arches are relatively very small. The same holds for the foundation and the abutments up to a certain height. For these structural parts, according to oral tradition, special attention was paid for the quality of the stone and mortar to be employed.

7. On the contrary, neither the shape nor the quality of the stones or the mortar was of equal importance for the mandrel walls. As can be seen in **Figure 10g** that depicts the remaining part of the Plaka Bridge, these mandrel walls were internally constructed with some form of rubble. However, in order to protect these parts from the weather conditions, the mandrel walls were also encased within facades of good-quality stone masonry (**Figure 10g**).

8. Because the primary and secondary main arches were constructed at different construction stages, there is a continuous cylindrical joint that lies between them (see **Figure 10a–f**). As revealed by the remains of the collapsed Plaka Bridge, wooden beams with iron inserts were employed to connect the primary and secondary arches at certain intervals.

9. Iron ties were also used to connect the two opposite faces of the primary arch in many bridges. These iron ties are visible in the photos of the main central arch of the Plaka Bridge before its collapse and they are still in place at the parts of the arch that were salvaged after its collapse (**Figures 11** and **12**). The iron ties were also used to connect the opposite faces of the primary arch of the main span in Tsipianis Bridge (**Figure 13**) and in Voidomatis bridge at Klidonia (**Figure 14**).

Figure 10. (a) Konitsa Bridge with the primary and secondary arches of the central span; (b) Plaka Bridge with the primary and secondary arches of the central span; (c) and (d) Kokorou Bridge with the primary and secondary arches of the central span; (e) and (f) Kontodimou Bridge with the primary and secondary arches of the central span; and (g) stone-masonry construction visible for the internal part of the mandrel walls of the remaining parts of Plaka Bridge.

Figure 11. Connection with wooden beams and iron inserts between the primary and secondary arches of the Plaka Bridge.

Figure 12. Iron ties used to connect the two opposite faces of the primary arch in Plaka Bridge.

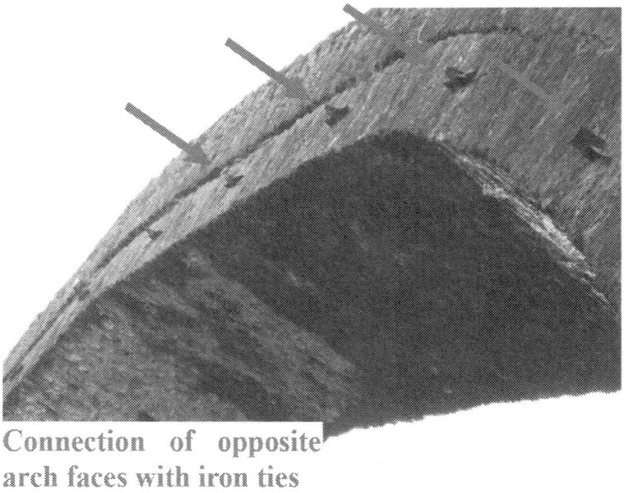

Figure 13. Iron ties used to connect the two opposite faces of the primary arch in Tsipianis Bridge.

Figure 14. Iron ties used to connect the two opposite faces of the primary arch in Voidomatis Bridge.

4. *In situ* measurements of the dynamic characteristics of stone bridges

4.1. Four studied stone bridges

In measuring the dynamic response of four stone bridges, two types of excitation were mobilized. The first, namely ambient excitation, mobilized the wind, despite the variation of the wind velocity in amplitude and orientation during the various tests. Due to the topography of the areas where these stone bridges are located, usually a relatively narrow gorge, the orientation of the wind resulted in a considerable component perpendicular to the longitudinal bridge axis (**Figures 15a**, **17a**, **18a** and **19a**). This fact combined with the resistance offered to this wind component by the façade of each bridge produced sufficient excitation source resulting in small amplitude vibrations that could be recorded by the employed instrumentation. For this purpose, the employed SysCom triaxial velocity sensors had a sensitivity of 0.001 mm/s and a SysCom data acquisition system with a sampling frequency of 400 Hz. All the obtained data were subsequently studied in the frequency domain through available fast Fourier transform (FFT) software [4, 5]. This wind orientation relative to the geometry of each bridge structure coupled with the bridge stiffness properties could excite mainly the first symmetric out-of-plane eigen-mode, as can be seen in **Figure 16c** for the Konitsa Bridge. The variability of the wind orientation could also excite, although to a lesser extent, some of the other in-plane and out-of-plane eigen-modes (see **Figure 16c** for the Konitsa Bridge).

The second type of excitation that was employed, namely vertical in-plane excitation, was produced from a sudden drop of a weight on the deck of each stone-masonry bridge [6, 7]. This weight was of the order of approximately 2.0 kN that was dropped from a relatively small height of 100 mm, so as to avoid even the slightest damage to the stone surface of the deck of each bridge. Again, the level of this second type of excitation was capable of producing mainly vertical vibrations and exciting the in-plane eigen-modes of each structure that could be captured by the employed SysCom triaxial velocity sensors with a sensitivity of 0.001 mm/s and a SysCom data acquisition system with a sampling frequency of 400 Hz. All the obtained data were subsequently studied in the frequency domain through available FFT software. In **Figure 15c**, the velocity measurements are depicted along the three axes (*x-x* horizontal out-of-plane, *y-y* horizontal in-plane and *z-z* vertical) as they were recorded during a typical sampling with the wind excitation. In **Figure 15d**, the velocity measurements are again depicted along the three axes (*x-x* horizontal out-of-plane, *y-y* horizontal in-plane and *z-z* vertical) as they were recorded during a typical sampling with the drop weight excitation. As can be seen, the drop weight excitation could produce at the dominant frequencies vibrations at least one order of magnitude larger than the wind excitation. From these measurements, an attempt was also made to obtain an estimate of the damping ratio for the dominant in-plane and out-of-plane frequencies. As is depicted in **Figure 16a** for the wind excitation, the main symmetric out-of-plane vibration that is excited by the wind has a dominant period of 2.539 Hz and a corresponding damping ratio approximately 1.7%. Similarly, as is depicted in **Figure**

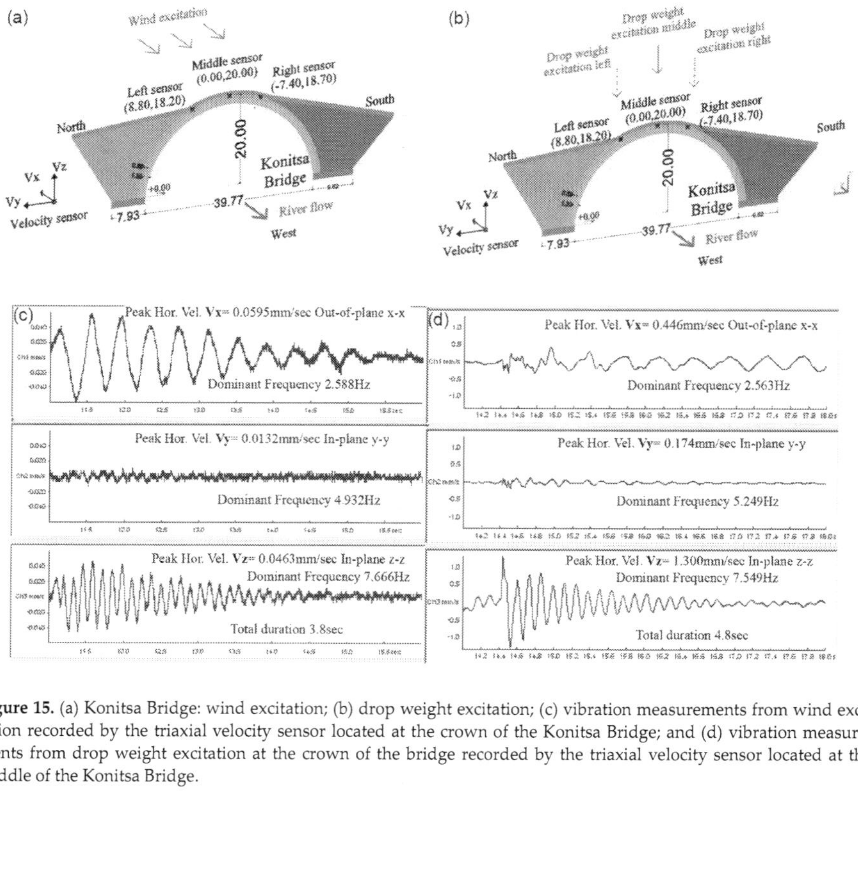

Figure 15. (a) Konitsa Bridge: wind excitation; (b) drop weight excitation; (c) vibration measurements from wind excitation recorded by the triaxial velocity sensor located at the crown of the Konitsa Bridge; and (d) vibration measurements from drop weight excitation at the crown of the bridge recorded by the triaxial velocity sensor located at the middle of the Konitsa Bridge.

16b for the drop weight excitation, the main symmetric in-plane vibration that is excited by the drop weight has a dominant period of 7.715 Hz and a corresponding damping ratio approximately 2.7%. This increase in the damping ratio value for this latter dominant frequency must be attributed to the relatively larger amplitudes of vibration that are produced from the drop weight excitation than from the wind excitation, as already underlined. All vibration measurements of the dynamic response of the Konitsa Bridge for either type of excitation were utilized to extract the eigen-frequencies depicted in **Figure 16c** together with the approximate shape of the corresponding eigen-modes.

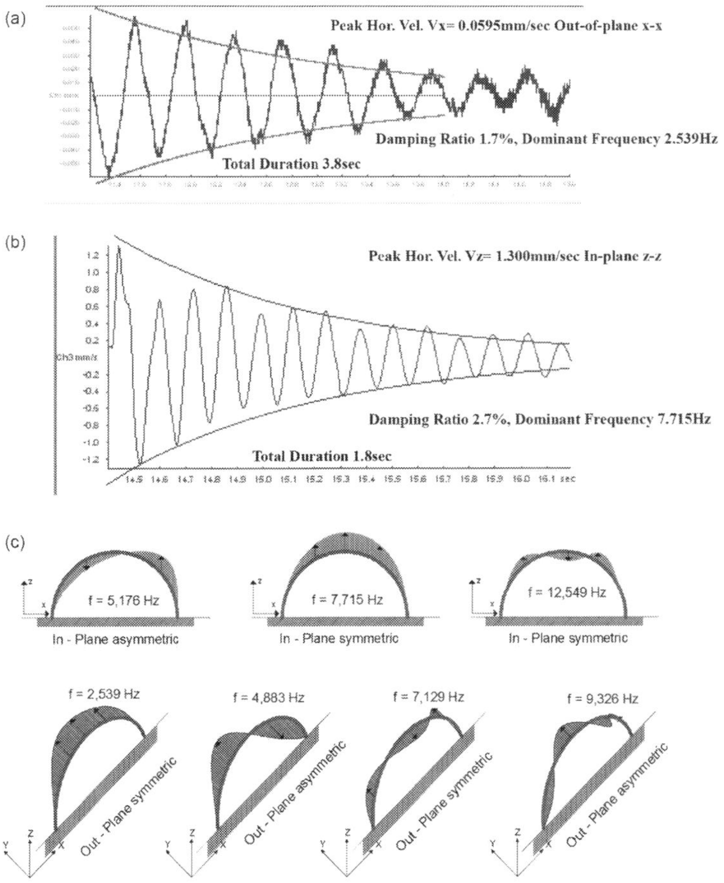

Figure 16. (a) Vibration measurements from wind excitation obtained from the triaxial velocity sensor located at the middle of the Konitsa Bridge; (b) vibration measurements from drop weight excitation at the middle of the bridge obtained from the triaxial velocity sensor also located at the middle of the Konitsa Bridge; and (c) measured eigen-frequencies and corresponding eigen-modes for the Konitsa Bridge.

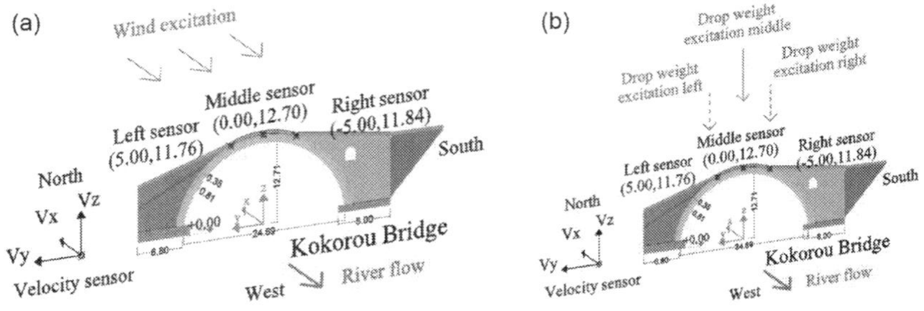

Figure 17. Kokorou Bridge: (a) wind excitation and (b) drop weight excitation.

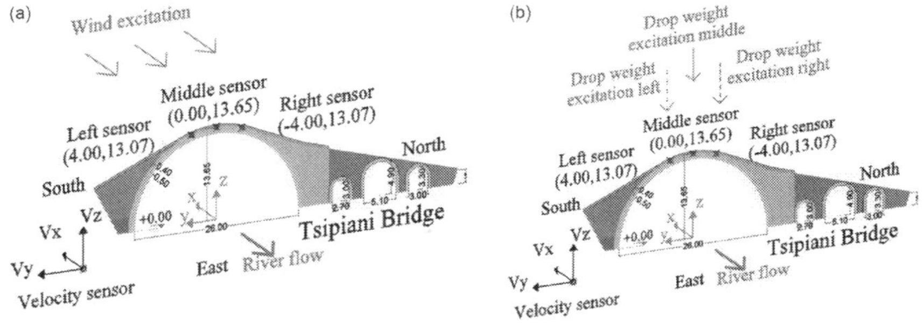

Figure 18. Tsipianis Bridge: (a) wind excitation and (b) drop weight excitation.

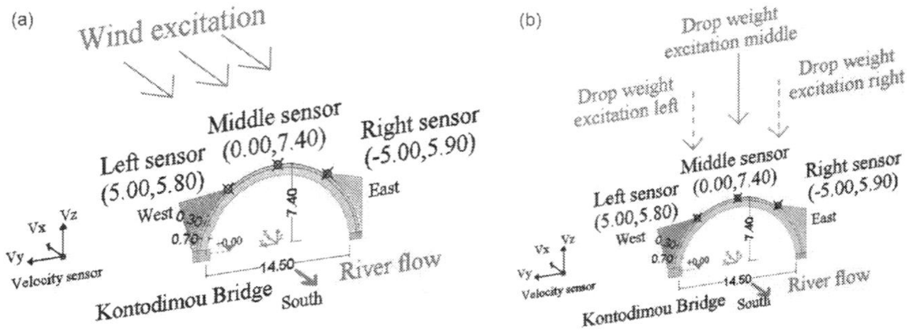

Figure 19. Kontodimou Bridge: (a) wind excitation and (b) drop weight excitation.

Measured eigen-frequencies (Hz) for the *Konitsa Bridge*			
In-plane	First asymmetric 5.176 Hz	Second symmetric 7.715 Hz	Third symmetric 12.549 Hz
Out-of plane	First symmetric 2.539 Hz	Second asymmetric 4.883 Hz	Third symmetric 7.129 Hz
Measured eigen-frequencies (Hz) for the *Kokorou Bridge*			
In-plane	First asymmetric 7.275 Hz	Second symmetric 10.059 Hz	Third symmetric 17.139 Hz
Out-of plane	First symmetric 4.541 Hz	Second asymmetric 7.471 Hz	Third symmetric 10.303 Hz
Measured eigen-frequencies (Hz) for the *Tsipiani Bridge*			
In-plane	First asymmetric 6.934 Hz	Second symmetric 8.549 Hz	Third symmetric 14.209 Hz
Out-of plane	First symmetric 3.320 Hz	Second asymmetric 6.348 Hz	Third symmetric 12.207 Hz
Measured eigen-frequencies (Hz) for the *Kontodimou Bridge*			
In-plane	First asymmetric 11.621 Hz	Second symmetric 16.934 Hz	Third symmetric 22.00 Hz
Out-of plane	First symmetric 7.860 Hz	Second asymmetric 16.846 Hz	Third symmetric 24.023 Hz

Table 1. Measured eigen-frequencies for four stone-masonry bridges.

The same process was followed for measuring the dynamic characteristics of another three stone-masonry bridges (Kokorou, Tsipianis and Kontodimou) using both the wind and the drop weight excitations, as shown in **Figures 17–19** where the position of the employed velocity sensors is indicated. Next, by utilizing all these vibration measurements of the dynamic response of each of these studied bridges for either type of excitation, it was possible to extract the relevant eigen-frequencies that are listed in **Table 1**. At least measurements of three repetitive sampling sequences for each type of excitation, either wind or drop weight, for each bridge (Konitsa, Kokorou, Tsipianis and Kontodimou) were measured. The eigen-frequency values listed in **Table 1** are values representing an average from corresponding values that were obtained by analysing the measured response from all tests.

4.2. Additional field measurements for the Konitsa Bridge

To gain more confidence in the *in situ* measurements presented in Section 4.1, the results of an independent *in situ* campaign are also presented here and briefly compared with the corresponding results presented in Section 4.1 for the same bridge. This additional *in situ* campaign was conducted during the end of October 2015 (**Figure 20**). This almost coincides with the *in situ* campaign described in Section 4.1, which was conducted during the period from mid-November 2015 till mid of December 2015 for all four bridges. Moreover, for the Konitsa Bridge the measurements presented in Section 4.1 were obtained on the dates of 8, 16 and 20 November. Based on this timing and the constant weather conditions prevailing during this period, no influence is expected to arise from environmental conditions to all these measurements. The objective of this independent field experiment was the same, that is, to assess the dynamic characteristics of the Konitsa stone arch bridge [8] using a set of Wilcoxon high-sensitivity accelerometers (1000 V/g) integrated with a data-recording/FFT analyzer RION-S78 system depicted in **Figure 20**. Further data post-processing was performed using a set of additional FFT processing software.

Figure 20. RION system and accelerometer system used in Konitsa Bridge field study. (a) Location of sensor at the South part of Konitsa Bridge and (b) location of the sensor at the North part of Konitsa Bridge.

Shown in **Figure 21** are the post-wind gust bridge response (vertical acceleration) and the corresponding power spectrum associated with the trace segment between 12 and 16 s of the record [4]. The power spectrum associated with the decay segment clearly delineates (a) the symmetric vertical mode (7.75 Hz).

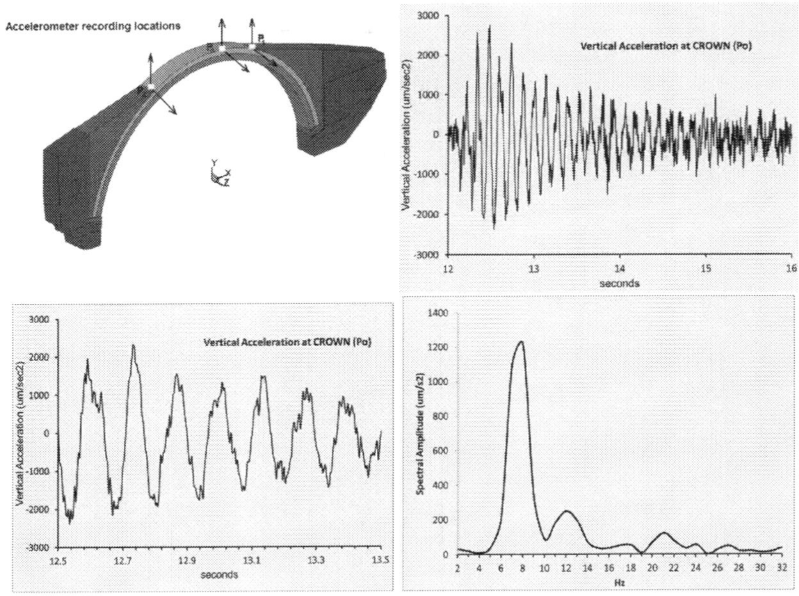

Figure 21. Decay segment of acceleration trace (vertical) and the corresponding power spectrum (7.75 Hz, damping ratio estimate of 1.6%).

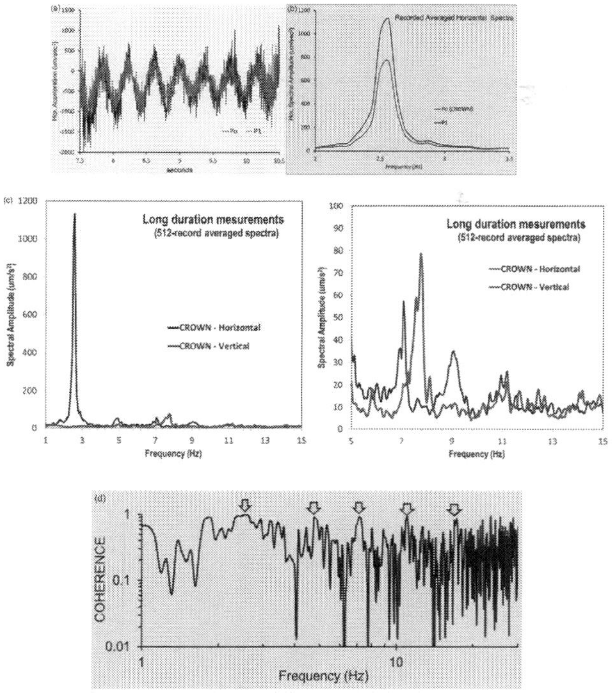

Figure 22. Horizontal (out-of-plane) acceleration time histories recorded simultaneously at two locations on the Konitsa Bridge deck. (a) Entire trace including high wind effects; (b) post-wind free vibration. Dominant frequency of 2.56 Hz; (c) recorded vertical and horizontal spectra at crown averaged over 512 records; (d) coherence measurements between P_0 and P_1 locations aiding mode identification (arrows indicate 100% coherence characteristic of the structure modes).

Figure 22 depicts the horizontal (out-of-plane) acceleration time histories recorded simultaneously at two locations on the Konitsa Bridge deck and their corresponding power spectrum with dominant frequency of 2.56 Hz (first out-out-of plane eigen-mode). Comparing the eigen-frequency values obtained for the *in situ* experiments, reported in Section 4.1 (depicted in **Figures 15** and **16**), with the corresponding values obtained from this independent *in situ* experiments (depicted in **Figures 21** and **22**), very good consistency can be observed. **Figure 22c** and **b** depict the FFT-averaged Fourier spectral curves that formed the base together with the coherence plot of **Figure 22d** to identify with confidence the eigen-frequency values [9].

5. Laboratory tests for the stone masonry

A laboratory testing sequence was performed having as an objective to study in a preliminary way the mechanical characteristics of the basic materials representative of the materials employed to build the studied stone-masonry bridges [4]. For this purpose, stone samples were selected from the neighbourhood of the collapsed Plaka Bridge as well as from a quarry near the Kontodimou and Kokorou Bridges. Moreover, stone samples were also taken from the river bed of the Kontodimou Bridge. Furthermore, it was possible to take a mortar sample from the collapsed Plaka Bridge. From both the stone and mortar samples collected *in situ*, it was possible to form specimens of regular prismatic geometry. These specimens were subjected to either axial compression or four-point bending tests. For the compression tests, the loaded surfaces of the prisms were properly cupped. **Figure 23a** and **c** depict typical loading arrangements employed for the compression (stone and mortar specimens) tests, whereas **Figure 23b** depicts the loading arrangement employed for the four-point bending tests. The applied load was measured through a load cell and the deformation of the tested specimens was measured employing a combination of displacement sensors as well as a number of strain gauges. These measurements were continuously recorded with a sampling frequency of 10 Hz. Through these measurements, the mechanical characteristics of the tested specimens were obtained in terms of compressive strength, flexural tensile strength, Young's modulus of elasticity and Poisson's ratio. The obtained values of these mechanical parameters are listed in **Tables 2–6**.

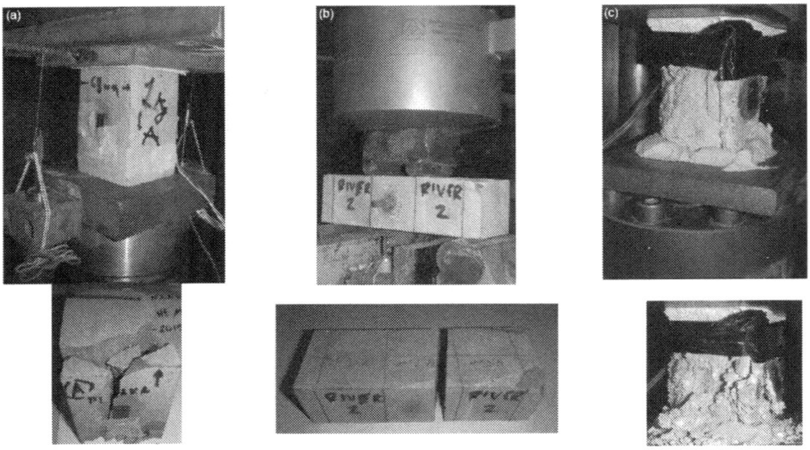

Figure 23. (a) Testing in compression stone samples taken from Plaka Bridge; (b) testing in four-point flexure stone sample taken from the river bed of Kontodimou Bridge; and (c) testing in compression mortar samples taken from Plaka Bridge.

Code name of sample	Cross section (mm^2)	Height (mm)	Maximum load (KN)	Compressive strength (MPa)	Slenderness ratio*/correction coefficient	Compressive strength (MPa) with correction due to slenderness*
River 1a	58.5 × 48.5	74.0	310.2	124.0	1.383/0.82	101.7
River1b	61.5 × 48.3	61.0	230.5	77.6	1.111/0.70	54.3

Code name of sample	Cross section (mm^2)	Height (mm)	Maximum load (KN)	Compressive strength (MPa)	Slenderness ratio*/correction coefficient	Compressive strength (MPa) with correction due to slenderness*
River 2a	56.3 × 45.0	59.5	225.6	89.0	1.175/0.73	65.0
River 2b	59.0 × 45.0	65.5	363.0	136.0	1.260/0.77	104.7
River 2c	33.0 × 65.0	59.0	220.7	102.9	1.205/0.75	77.2
Kontodimou Bridge River stone			**Average compressive strength** = 80.6 MPa, E_1 = 55,560 MPa, ν = 0.259**			
Quarry 1a	55.0 × 45.5	70.0	416.9	166.6	1.394/0.82	136.6
Quarry 1b	52.0 × 46.0	73.5	230.5	96.4	1.500/0.90	88.8
Quarry 2a	44.8 × 44.8	84.5	193.3	96.3	1.886/0.95	91.5
Quarry 2b	47.0 × 44.8	43.5	313.9	149.0	0.948/0.65	96.9
Kontodimou Bridge Quarry stone			**Average compressive strength** = 103.5 MPa, E_1 = 85,000 MPa, ν = 0.3**			

*Reference slenderness ratio = 2.0.
**The average compressive strength refers to a prism with a slenderness ratio = 2.

Table 2. Compression tests (22 January 2016) with stone samples taken near Kontodimou Bridge.

Code name of sample	Width (mm)	Height (mm)	Span (mm)	Maximum vertical load (kN)	Tensile strength (MPa)	Young's Modulus from flexure (MPa) (DCDT)
River 1	60.0	46.5	135.0	24.98	23.41	2875
River 2	60.0	45.0	135.0	22.66	25.18	1545
Kontodimou Bridge River stone			**Average tensile flex. strength = 24.30 MPa, E_2 = 2210 MPa**			
Quarry 1	54.0	46.5	135.0	12.875	14.89	11,205
Quarry 2	45.0	44.5	135.0	13.647	20.67	15,345
Kontodimou Bridge Quarry stone			**Average tensile flex. strength = 17.78 MPa, E_2 = 13,275 MPa**			

Table 3. Flexure tests (15 January 2016) with stone samples taken near Kontodimou Bridge.

Code name of sample	Cross section (mm^2)	Height (mm)	Maximum load (kN)	Compressive strength (MPa)	Slenderness ratio*/ correction coefficient	Compressive strength (MPa) with correction due to slenderness*
Specimen A	61.0 × 68.0	93.0	264.9	63.9	1.442/0.87	55.6
Specimen B	67.5 × 62.0	89.0	443.4	106.0	1.375/0.82	86.9
Plaka Bridge stone Specimens			**Average compressive strength** = 71.3 MPa, E_1 = 40,000 MPa, ν = 0.142**			

*Reference slenderness ratio = 2.0.
**The average compressive strength refers to a prism with a slenderness ratio = 2.

Table 4. Compression tests (18 December 2015) with stone samples taken at Plaka Bridge.

Code name of sample	Width (mm)	Height (mm)	Span (mm)	Maximum vertical load (kN)	Tensile strength (MPa)	Young's Modulus from flexure (MPa) (S.G.)
Specimen A	52.0	52.0	180.0	14.75	18.88	33,330
Specimen B	52.0	52.0	180.0	12.13	15.52	36,360
Plaka Bridge stone specimens			**Average tensile flex. strength = 17.20 MPa, E_2 = 34,845 MPa**			

Table 5. Flexure tests (18 December 2015) with stone samples taken at Plaka Bridge.

Code name of sample	Cross section (mm^2)	Height (mm)	Maximum load (kN)	Compressive strength (MPa)	Slenderness ratio*/ correction coefficient	Compressive strength (MPa) with correction due to slenderness*
Specimen 1	27.5 × 57.0	66.0	3.228	2.06	1.562/0.91	1.875
Plaka Bridge stone specimens			**Compressive strength = 1.875 MPa, E_1 = 2500 MPa, ν = 0.35**			

Table 6. Compression tests (28 January 2016) with mortar samples taken at Plaka Bridge.

6. Numerical simulation of dynamic characteristics

In this section, the dynamic characteristics of the four studied stone-masonry bridges will be predicted through a numerical simulation process. Initially, this numerical simulation will be based on elastic behaviour, assuming the stone masonry as an orthotropic continuous medium and limiting these numerical models at approximately the interface between the end abutments and the rocky river banks, thus introducing boundaries at these locations [10]. For simplicity purposes, the bulk of these numerical simulations are made in the 3-D domain representing these bridge structures with their mid-surface employing thick-shell finite elements [11]. The various main parts of these stone-masonry bridges, that is, the primary and the secondary arches, the abutments, the deck, the mandrel walls and the parapets, were simulated in such a way that narrow contact surfaces could be introduced between them, representing in this way a different 'softer' medium. All available information, measured during the *in situ* campaign, on the geometry of each one of these parts for every bridge was used in building up these numerical simulations. The mechanical property values obtained from the stone and mortar sample tests, which were presented in Section 5, indicate the following main points. Young's modulus of the stone samples in axial compression has a value exceeding 40 GPa, whereas they yield a much less stiff behaviour in flexure. It is well known that the complex triaxial behaviour of masonry cannot be easily approximated from the mechanical behaviour of its constituents. For the studied stone-masonry bridges, this becomes even more difficult considering the various construction stages that were discussed in Section 3, the variability of the materials employed to form the distinct parts during these construction stages and the interconnection and contact conditions between the various parts formed during these construction stages (abutments, primary and secondary arches, deck, parapets, mandrel walls). Moreover, there is important information that is needed in order to form with some realism the boundary conditions at the river bed and banks [11]. The lack of specific studies towards clarifying in a systematic way all these uncertainties represents a serious limitation in the numerical simulation process.

Measured/predicted eigen-frequencies (Hz) for the ***Konitsa Bridge***			
In-plane	First asymmetric 5.176/6.724	Second symmetric 7.715/7.076	Third symmetric 12.549/10.065
Out-of plane	First symmetric 2.539/2.432	Second asymmetric 4.883/4.478	Third symmetric 7.129/7.275
Measured eigen-frequencies (Hz) for the ***Kokorou Bridge***			
In-plane	First asymmetric 7.275/10.212	Second symmetric 10.059/11.134	Third symmetric 17.139/15.125
Out-of plane	First symmetric 4.541/4.035	Second asymmetric 7.471/7.065	Third symmetric 10.303/11.545
Measured eigen-frequencies (Hz) for the ***Tsipiani Bridge***			
In-plane	First asymmetric 6.934/7.106	Second symmetric 8.549/10.029	Third symmetric 14.209/12.742
Out-of plane	First symmetric 3.320/3.090	Second asymmetric 6.348/5.734	Third symmetric 12.207/9.248
Measured eigen-frequencies (Hz) for the ***Kontodimou Bridge***			
In-plane	First asymmetric 11.621/11.893	Second symmetric 16.934/12.389	Third symmetric 22.000/20.155
Out-of plane	First symmetric 7.860/5.664	Second asymmetric 16.846/13.117	Third symmetric 24.023/21.988

Table 7. Comparison of ***measured/predicted*** eigen-frequencies for four stone-masonry bridges (***pinned*** boundary conditions).

Measured/predicted eigen-frequencies (Hz) for the ***Konitsa Bridge***			
In-plane	First asymmetric 5.176/6.733	Second symmetric 7.715/7.078	Third symmetric 12.549/10.085
Out-of plane	First symmetric 2.539/2.526	Second asymmetric 4.883/4.759	Third symmetric 7.129/7.706
Measured eigen-frequencies (Hz) for the ***Kokorou Bridge***			
In-plane	First asymmetric 7.275/10.212	Second symmetric 10.059/11.134	Third symmetric 17.139/15.125
Out-of plane	First symmetric 4.541/4.548	Second asymmetric 7.471/8.408	Third symmetric 10.303/13.733
Measured eigen-frequencies (Hz) for the ***Tsipiani Bridge***			
In-plane	First asymmetric 6.934/7.106	Second symmetric 8.549/10.029	Third symmetric 14.209/12.742
Out-of plane	First symmetric 3.320/3.321	Second asymmetric 6.348/6.257	Third symmetric 12.207/10.012
Measured eigen-frequencies (Hz) for the ***Kontodimou Bridge***			
In-plane	First Asymmetric 11.621/11.905	Second Symmetric 16.934/12.395	Third Symmetric 22.000/20.185
Out-of plane	First Symmetric 7.860/7.643	Second Asymmetric 16.846/17.035	Third Symmetric 24.023/26.362

Konitsa Bridge. Emasonry = 4000 MPa, Econtact = 2000 MPa. Bending Stiffness Modifiers = 3.0.
Kokorou Bridge. Emasonry = 4000 MPa, Econtact = 2000 MPa. Bending Stiffness Modifiers = 1.75.
Tsipianis Bridge. Emasonry = 4000 MPa, Econtact = 2000 MPa. Bending Stiffness Modifiers = 1.0.
Kontodimou Bridge. Emasonry = 1600 MPa, Econtact = 1600 MPa. Bending Stiffness Modifiers = 1.0.

Table 8. Comparison of ***measured/predicted*** eigen-frequencies for four stone-masonry bridges (***fixed*** boundary conditions).

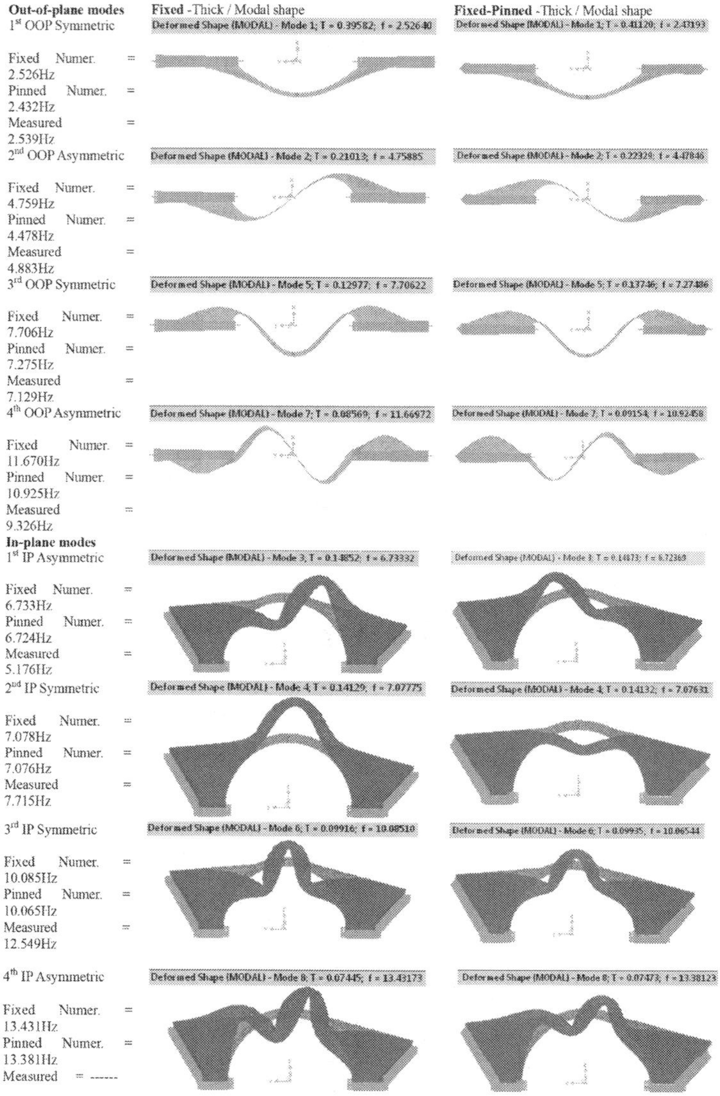

Figure 24. Numerical and observed eigen-values for the ***Konitsa Bridge***. $E_{masonry}$ =4000MPa, $E_{contact}$ =2000MPa. Bending Stiffness Modifiers = 3.0.

The approximation adopted in this study is a process of back simulation [6, 7]. That is, adopting values for these unknown mechanical stone-masonry properties, respecting at the same time all the measured geometric details, which result in reasonably good agreement between the measured and predicted in this way eigen-frequency values. Following this approximate process, two distinct cases of boundary conditions were introduced. In one series of numerical simulations, all the boundaries, either at the river bed or at the river banks, were considered as being fixed in these 3-D numerical simulations for all studied bridges. This is denoted in the predicted eigen-frequency values in **Tables 7** and **8** and **Figures 24** and **25** with the subscript '*Fixed Numer*'. Alternatively, the rotational degrees of freedom were released all along the locations where the abutments are supported at the river banks thus excluding the footings. This is denoted in the predicted eigen-frequency values in **Tables 7** and **8** and **Figures 24** and **25** with the subscript '*Pinned Numer*'. It is shown from this sensitivity analysis that this variation in the boundary conditions approximation influences, as expected, the out-of-plane and not the in-plane stiffness of the studied stone-masonry bridges. This out-of-plane stiffness variation is more pronounced for the relatively small dimensions Kontodimou Bridge rather than for the relatively large Konitsa Bridge and Plaka Bridge. Moreover, for the Tsipianis Bridge

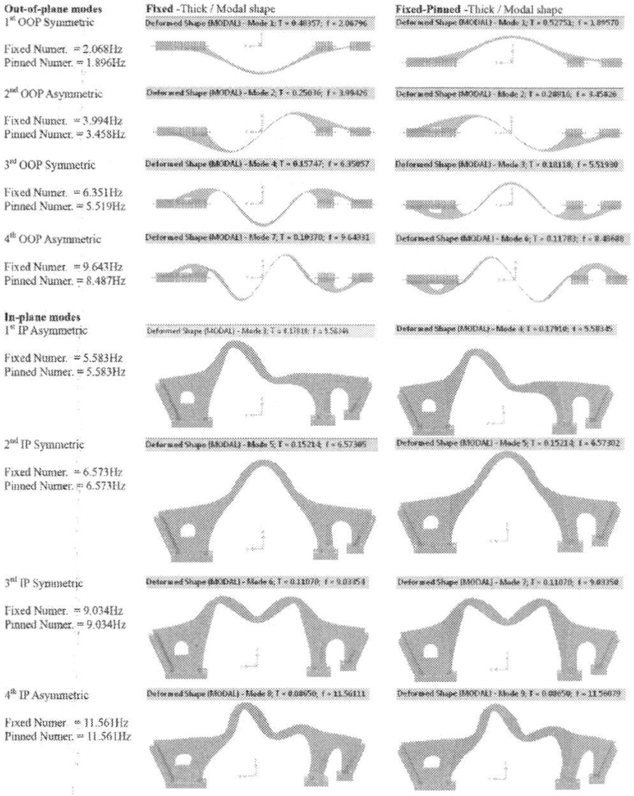

Figure 25. Numerical eigen-values for the Plaka Bridge. ***Plaka Bridge***. $E_{masonry}$ = 4000 MPa, $E_{contact}$ = 2000 MPa. Bending stiffness modifiers = 3.0.

whereby the main central arch is supported at the North end in adjacent arches rather than on the rocky bank, this variation of the boundary conditions, as expected, has again a less pronounced influence. The value of Young's modulus that was adopted for the masonry in these numerical simulations is listed at the bottom of **Table 8** and at the captions of **Figures 24** and **25** that depict the numerical eigen-mode and eigen-frequency numerical results together with the measured *in situ* eigen-frequency values for each bridge. Thus, for the Konitsa (**Figure 24**), Kokorou and Tsipianis Bridges, 4 GPa was adopted for the masonry Young's modulus and 2 GPa for the contact surface. For the Kontodimou Bridge, these values were 1.6 GPa for both the masonry and the contact surface. A partial explanation is that the mortar joints and contact surface between the various bridge parts in the Kontodimou Bridge (**Figure 10e** and **f**) were wider than in other bridges and the mortar was in some cases washed out at some depth. In order to approximate the in-plane and the out-of-plane stiffness of the studied stone-masonry bridges, which directly influences the corresponding numerical eigen-frequency values, listed in **Tables 7** and **8** and depicted in **Figures 24** and **25**, a flexural stiffness amplifier was introduced for the Konitsa Bridge and the Kokorou Bridge equal to 3.0 and 1.75, respectively. From the comparison of the results of these numerical simulations in terms of eigen-frequencies and eigen-modes, listed in **Tables 7** and **8** and depicted in **Figures 24** and **25**, it can be seen that in most cases the predicted eigen-frequency values are in reasonably good agreement with the measured values. Moreover, the order of the out-of-plane and the in-plane eigen-modes predicted by the numerical simulation is in agreement with the observed response. An exception is the first asymmetric in-plane eigen-mode for the Konitsa Bridge (**Figure 24**) and Kokorou Bridge (**Tables 7** and **8**) that indicates a corresponding measured stiffness smaller than the predicted one. On the basis of this comparison, an additional numerical simulation was performed for the Plaka Bridge (**Figure 25**), despite the lack of measured response in this case, adopting the same assumptions that were described before specifically for the Konitsa Bridge. As can be seen by comparing the numerical eigen-frequency values of the Konitsa Bridge (**Figure 24**) with those of the Plaka Bridge (**Figure 25**), the latter, as expected, is more flexible both in the in-plane and in the out-of-plane direction.

7. Simplified numerical investigation of the seismic behaviour of the studied stone-masonry bridges

7.1. Simplified dynamic spectral numerical simulation of the seismic behaviour of the Konitsa Bridge

This section includes results of a series of numerical simulations of the Konitsa Bridge when it is subjected to a combination of actions that include the dead weight (D) combined with seismic forces. The seismic forces will be defined in various ways, as will be described in what follows. Initially, use is made of the current definition of the seismic forces by EURO-Code 8 [12]. Towards this, horizontal and vertical design spectral curves are derived based on the horizontal design ground acceleration. This value, as defined by the zoning map of the current Seismic Code of Greece [13, 14], is equal to 0.16 g (g is the acceleration of gravity) for the location

http://dx.doi.org/10.5772/64752

of the Konitsa Bridge. Furthermore, it is assumed that the soil conditions belong to category A because of the rocky site where this bridge is founded, that the importance and foundation coefficients have values equal to one (1.0); the damping ratio is considered equal to 5% and the behaviour factor is equal to 1.5 (unreinforced masonry). The design acceleration spectral curves obtained in this way are depicted in **Figure 26a** and **b** for the horizontal and vertical direction, respectively. In the same figures, the corresponding elastic acceleration spectral curves are also shown derived from the ground acceleration recorded during the main event of the earthquake sequence of 5 August 1996 at the city of Konitsa located at a distance of approximately 1.5 km from the site of the bridge [15]. In **Figure 26a** and **b**, the eigen-period range of the first 12 eigen-modes is also indicated (ranging between the low and the high modal period). For the vertical response spectra, this is done for only the in-plane eigen-modes (see also **Table 9**).

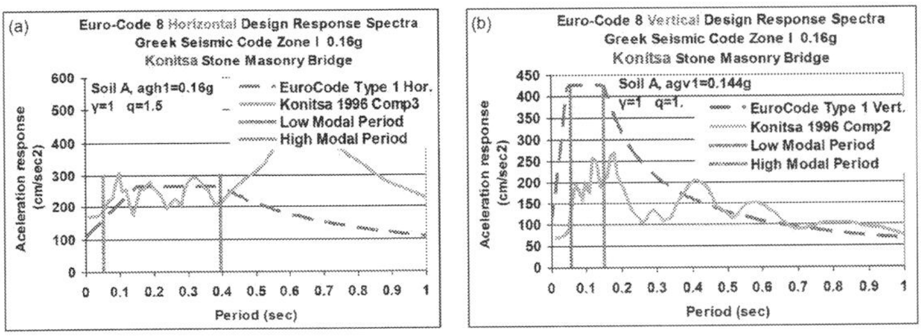

Figure 26. (a) Horizontal spectral curves for the 1996-Konitsa earthquake and the type-1 Euro-Code and (b) vertical spectral curves for the 1996-Konitsa earthquake and the type-1 Euro-Code.

As can be seen in **Figure 26a**, the Euro-Code horizontal acceleration spectral curves compare well with the horizontal component-3 of 1996 Earthquake spectral curves for the period range of interest. The Euro-Code vertical acceleration spectral curves, depicted in **Figure 26b**, are approximately 100% larger than the vertical component-2 of 1996 Earthquake spectral curves for the period range of interest. Based on these plots, it can be concluded that this bridge sustained a ground motion that in the horizontal direction was approximately comparable to the design earthquake; however, the design earthquake in the vertical direction is shown to be more severe than the one this stone-masonry bridge experienced during the 1996 earthquake sequence.

In **Table 10**, the base reactions are listed (F_X, F_Y and F_Z) in the x-x ($u1$, out-of-plane), the y-y ($u2$, in-plane) and z-z ($u3$, in-plane) directions (see **Figure 15a** and **b**) from the various load cases, which were considered in this numerical study. Apart from the dead load (D, row 1) in rows 2–4 of **Table 10**, the base reaction values listed are obtained from dynamic spectral analyses employing the horizontal and vertical response spectral curves of the 1996-Konitsa earthquake event (**Figure 26a** and **b**). In rows 7–9 of **Table 10**, the base reaction values are again obtained from dynamic spectral analyses employing this time the Euro-Code horizontal and vertical design spectral curves of **Figure 26a** and **b**. In all these dynamic spectral analyses, the 12 eigen-modes listed in **Table 9** were employed.

Output case	Period	Frequency	U_X	U_Y	U_Z	SumU_X	SumU_Y	SumU_Z
Text	s	Hz	Unitless	Unitless	Unitless	Unitless	Unitless	Unitless
Mode 1 (first OOP Symmetric)	0.396	2.5264	0.3300	0	0	0.33002	0	0
Mode 2 (second OOP asymmetric)	0.210	4.7588	0.0004	0	0	0.33043	0	0
Mode 3 (first IP asymmetric)	0.149	6.7333	0.0	0.08498	0.00055	0.33043	0.08498	0.00055
Mode 4 (second IP symmetric)	0.141	7.0777	0.0	0.00104	0.12769	0.33043	0.08602	0.12825
Mode 5 (third OOP symmetric)	0.130	7.7062	0.2271	0	0	0.55753	0.08602	0.12825
Mode 6 (third IP symmetric)	0.099	10.0851	0.0	0.0007	0.15393	0.55753	0.08672	0.28218
Mode 7 (fourth OOP asymmetric)	0.086	11.6697	0.0	0	0	0.55754	0.08672	0.28218
Mode 8 (fourth IP asymmetric)	0.074	13.4317	0.0	0.17916	0.0023	0.55754	0.26588	0.28447
Mode 9 (fifth OOP symmetric)	0.065	15.2999	0.08851	0	0	0.64605	0.26588	0.28447
Mode 10 (fifth IP symmetric)	0.062	16.0400	0.0	0.23382	0.000014	0.64605	0.49971	0.28449
Mode 11 (sixth IP asymmetric)	0.0565	17.6794	0.0	0.0026	0.08175	0.64605	0.5023	0.36624
Mode 12 (sixth OOP asymmetric)	0.0522	19.1439	0.00262	0	0	0.64868	0.5023	0.36624

Table 9. Modal participating mass ratios for Konitsa Bridge (see **Figure 24**).

As can be seen in **Table 9**, these eigen-modes have modal mass participation ratios that result in sums smaller than 90%. That is, SumU_x = 64.9%, SumU_y = 50.2% and SumU_z = 36.6% of the total mass for the direction of motion in the U_x, U_y and U_z axes, respectively. This was accounted for in the subsequent load combinations where the dead load is combined with the horizontal and vertical spectral curves (rows 5 and 6 of **Table 10**, Combination 1, 1996 earthquake horizontal + vertical spectral curves and rows 10 and 11 of **Table 10**, Combination 7 Euro-Code horizontal + vertical spectral curves). Towards this end, the dynamic spectral analysis results were multiplied by an amplification factor equal to the reverse of the relevant ratio values before superimposing the dead load results. This amplification factor is equal to 1/SumU_x for

http://dx.doi.org/10.5772/64752

the dynamic analyses employing the out-of-plane *x*-*x* horizontal eigen-modal ratio, to 1/SumU_y for the in-plane *y*-*y* horizontal eigen-modal ratios and to 1/SumU_z for the in-plane vertical eigen-modal ratios [11]. This becomes evident when one compares the base reaction values without and with these amplification factor values in **Table 10**.

	Loading case description	Loading type description	Type limit	Global F_X (kN)	Global F_Y (kN)	Global F_Z (kN)
1	DEAD (D)	Linear Static		0	0	35,853
2	1996 Comp 2 RS Ver IP	Linear Resp. Spectral 1996 EQ	Max	0	179	1499
3	1996 Comp 3 RS Hor *u*1 OP	Linear Resp. Spectral 1996 EQ	Max	3130	0	0
4	1996 Comp 3 RS Hor *u*2 IP	Linear Resp. Spectral 1996 EQ	Max	0	2552	230
5	Combination 1	Dead + 1996 EQ RS (*u*1+ *u*2 + *u*3)	Max	4825	5574	40,406
6	Combination 1	Dead + 1996 EQ RS (*u*1+ *u*2 + *u*3)	Min	−4825	−5574	31,299
7	Euro-Code RS Hor *u*1 OP	Linear Resp. Spectral Euro-Code	Max	3725	0	0
8	Euro-Code RS Hor *u*2 IP	Linear Resp. Spectral Euro-Code	Max	0	2224	193
9	Euro-Code RS Ver *u*3 IP	Linear Resp. Spectral Euro-Code	Max	0	443	3426
10	Combination 7	Dead + Euro-Code RS (*u*1+ *u*2 + *u*3)	Max	5742	5640	45,597
11	Combination 7	Dead + Euro-Code RS (*u*1+ *u*2 + *u*3)	Min	−5742	−5640	26,109

Table 10. Base reactions from the dynamic spectral analyses, Konitsa Bridge.

In **Figure 27a** and **b**, the numerically predicted deformation patterns of Konitsa Bridge are depicted for load combination 1 and 7, respectively. As can be seen, this stone-masonry bridge develops under these combinations of dead load and seismic forces relatively large out-of-plane displacements at the top of the main arch. As expected, the deformations for the Euro-Code design spectra reach the largest values attaining at the crown of the arch a maximum value equal to 30.6 mm. In **Figure 28a–h**, the numerically predicted state of stress (max/min S11, max/min S22), which develops at Konitsa Bridge for load combinations 1 and 7, is depicted. Again, as expected, the most demanding state of stress results for the load combination 7 that includes seismic forces provided by Euro-Code [12]. The largest values of tensile stress S11 (3.46 MPa, **Figure 28b**) develop at the bottom fibre of the crown of the arch. This is a relatively large tensile stress value that is expected to exceed the tensile capacity of the stone masonry of this bridge [16]. The largest value of tensile stress S22 (1.5 1 MPa, **Figure 28f**) develops at the area where the primary arch joins the foundation block. Again, this is a relatively large tensile stress value that is expected to exceed the tensile capacity of the stone masonry of this bridge. Both these remarks indicate locations of distress for this stone-masonry bridge predicting in this way the appearance of structural damage. On the contrary, the largest value of compression stress equal to S11 = −4.3 MPa (**Figure 28d**, for combination 7) is expected to be easily met by the compression capacity of the stone masonry for this bridge [16].

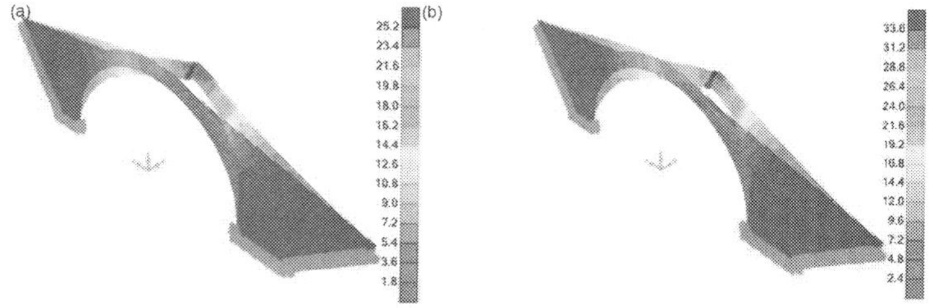

Figure 27. (a) Deformations of Konitsa Bridge. For loads Dead + 1996 EQ RS ($u1$+ $u2$ + $u3$). Comb 1. At crown $u1$ = 24.643 mm, $u2$ = 1.209 mm, $u3$ = −10.593 mm. (b) Deformations of Konitsa Bridge. For loads Dead + Euro-Code RS ($u1$+ $u2$ + $u3$). Comb 7. At crown $u1$ = 30.573 mm, $u2$ = 1.348 mm, $u3$ = −15.804 mm.

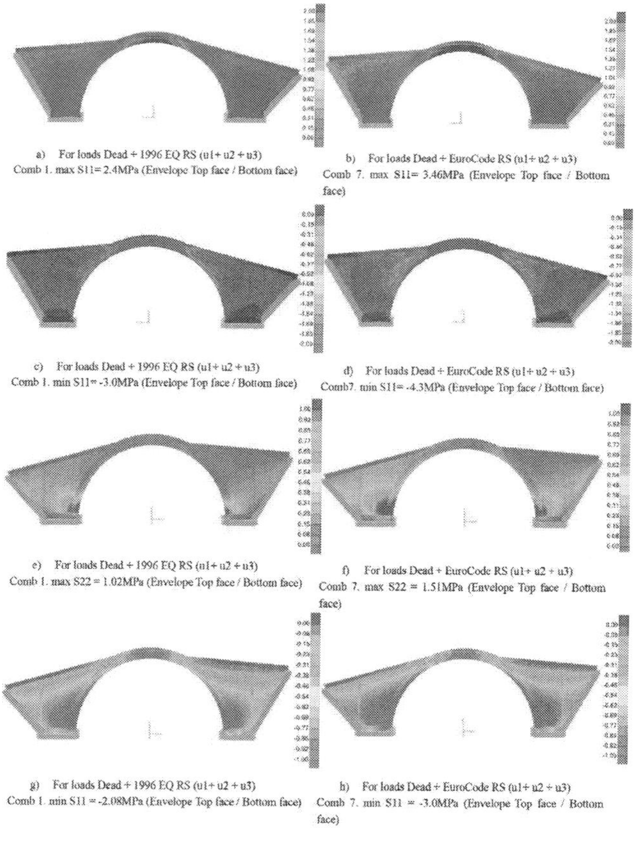

Figure 28. State of stress through the distribution of stresses S11 and S22 for Konitsa Bridge.

7.2. Dynamic elastic time-history numerical simulation of the seismic behaviour of the Konitsa Bridge

An additional linear numerical simulation was performed. This time, apart from the dead load (D, row 1, **Table 11**), the Konitsa Bridge was subjected to the horizontal component (Comp3) and/or the vertical component [17] of the 1996-Konitsa earthquake record (**Figure 29**) in the following way. The bridge was subjected only to the vertical (Comp2-Ez, rows 2 and 3 of **Table 11**) or only to the horizontal component of this record in the out of-plane direction (Comp3-Ex, rows 4 and 5 of **Table 11**). Alternatively, the bridge was subjected to the horizontal component of this record in the in-plane horizontal direction (Comp3-Ey, rows 6 and 7, **Table 11**) [15].

The solution this time was obtained through a step-by-step time integration scheme assuming a damping ratio equal to 5% of critical. In these analyses, only the first most intense 6 s of this 1996-Konitsa earthquake record were used [15]. In **Table 11**, the base shear values in the *x-x* (F_X, *u*1, horizontal out-of-plane), *y-y* (F_Y, *u*2, horizontal in-plane) and *z-z* (F_Z, *u*3, vertical) directions are listed in terms of limit values (maximum or minimum) that arose during the 6 s of these time-history analyses. Limit (maximum or minimum) base shear F_X, F_Y, F_Z values are also listed in rows 8–13 when these seismic excitations (Ex, Ey and Ez) are combined within themselves and the dead load as is shown in the third column of **Table 11** to produce load combinations encoded as COMB9, COMB10 and COMB11. By comparing these base shear values with the ones listed in **Table 10** where the response spectral curves of either the 1996-Konitsa record or the Euro-Code were employed, it can be seen that the limit (max/min) base shear amplitudes in both tables are very similar. **Figure 30a** shows the horizontal (u_x, out-of-plane) and the vertical (u_z, in-plane) displacement response at the crown of the Konitsa Bridge, obtained from the time-history numerical analyses. The horizontal response was obtained when the structure was subjected to horizontal component (Comp3) of the Konitsa 1996 earthquake record and the vertical in-plane response when the structure is subjected to vertical component (Comp2) of the Konitsa 1996 earthquake record (**Figure 29**). **Figure 30b** shows the variation of the S11 stress response at the bottom fibre of the crown of the Konitsa Bridge when this structure is subjected to either the horizontal component of the Konitsa 1996 earthquake record (Comp3) in the out-of-plane (u_x) direction or the vertical component of the Konitsa 1996 earthquake record (Comp2) in the vertical (u_z) in-plane direction. The location of the plotted stress is at the bottom fibre at the middle of the arch (crown) of the Konitsa Bridge. As can be seen in both **Figure 30a** and **b**, the horizontal u_x displacement and S11 stress response produced by the horizontal out-of-plane excitation are larger than the corresponding response vertical u_z displacement and S11 stress response produced by the vertical in-plane excitation. Moreover, as expected from the relevant response spectral curves depicted and the dominant eigen-frequency values (**Figures 24**, **26a** and **b**), the vertical u_z displacement and S11 stress response, produced by the vertical in-plane excitation, are of higher frequency content than the horizontal u_x displacement and S11 stress response produced by the horizontal out-of-plane excitation.

	Loading case description	Loading type description	Type limit	Global F_X (kN)	Global F_Y (kN)	Global F_Z (kN)
1	DEAD	Dead Load (D) Linear Static		0	0	35,853
2	Comp 2 TH Ver *u*3 IP	Konitsa 1996 Comp 2 THist. Ver *u*3 In-Plane (Ez)	Max	0	128.6	2651.2
3	Comp 2 TH Ver *u*3 IP	Konitsa 1996 Comp 2 THist. Ver *u*3 In-Plane (Ez)	Min	0	−140.2	−2380.5
4	Comp 3 TH Hor *u*1 OP	Konitsa 1996 Comp 3 THist. Hor *u*1 Out-of-Plane (Ex)	Max	5254.6	0	0
5	Comp 3 TH Hor *u*1 OP	Konitsa 1996 Comp 3 THist. Hor *u*1 Out-of-Plane (Ex)	Min	−5786.6	0	0
6	Comp 3 TH Hor *u*2 IP	Konitsa 1996 Comp 3 THist. Hor *u*2 In-Plane (Ey)	Max	0	4608.7	88.2
7	Comp 3 TH Hor *u*2 IP	Konitsa 1996 Comp 3 THist. Hor *u*2 In-Plane (Ey)	Min	0	−6456.7	−81.9
8	COMB9	Dead + Ex + Ez	Max	5254.6	128.6	38,504.2
9	COMB9	Dead + Ex + Ez	Min	−5786.6	−140.2	33,472.5
10	COMB10	Dead + Ey + Ez	Max	0	4737.3	38,592.4
11	COMB10	Dead + Ey + Ez	Min	0	-6596.9	33,390.7
12	COMB11	Dead + Ex + Ey + Ez	Max	5254.6	4737.3	38,592.4
13	COMB11	Dead + Ex + Ey + Ez	Min	−5786.6	−6596.9	33,390.7

Table 11. Base reactions from time-history analyses: Konitsa Bridge.

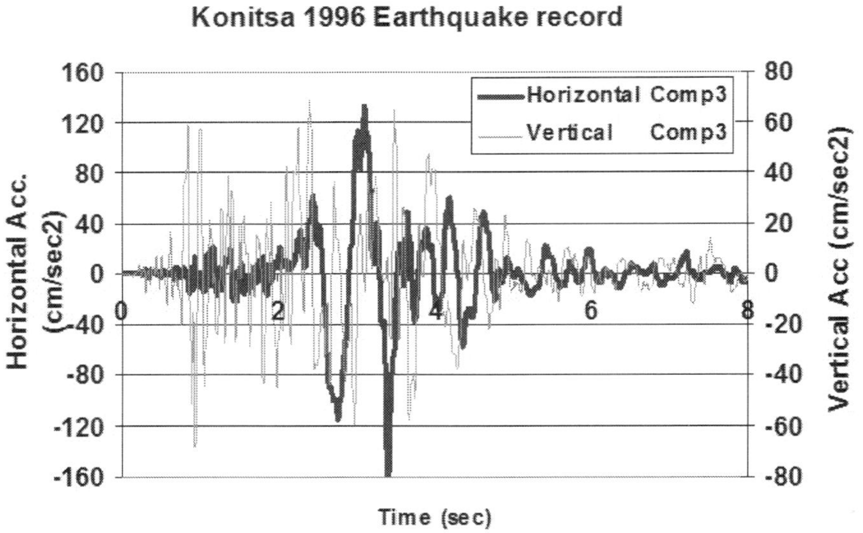

Figure 29. The first eight (8) most intense seconds of the 1996 earthquake record (ITSAK).

http://dx.doi.org/10.5772/64752

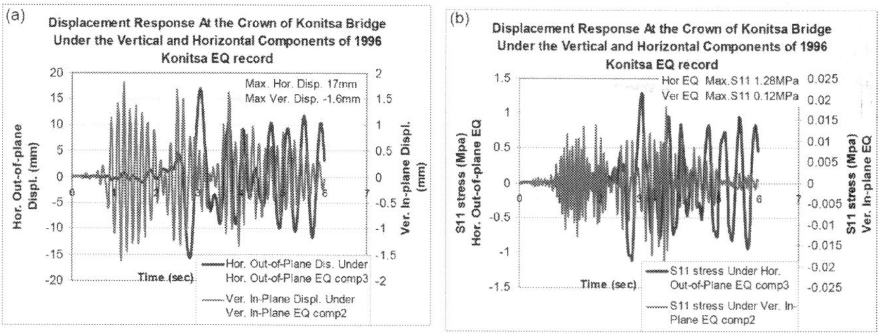

Figure 30. (a) Displacement (Hor. or Ver.) response at the crown of the Konitsa Bridge when subjected to either the horizontal or the vertical component of the Konitsa 1996 earthquake. (b) S11 stress response at the bottom of crown of the Konitsa Bridge when subjected to either the horizontal or the vertical component of the Konitsa 1996 earthquake.

Figure 31a and **b** depict the envelop of the limit (maximum/minimum) values of the S11 stress distribution in the Konitsa Bridge for load combination 11 that includes the dead load, the application of Comp3 of the Konitsa earthquake record in both the horizontal in-plane and out-of-plane direction as well as Comp2 of the Konitsa earthquake record in the vertical in-plane direction. By examining the displacement and stress response, it could be concluded that the application of the horizontal component of the Konitsa 1996 in the horizontal u_y in-plane direction is of too small amplitude to be of any significance. This must be attributed to the stiffness properties of this bridge in this direction and the resulting in-plane eigen-frequencies and eigen-modes that combined with the frequency content of this record result in displacement and stress response of relatively small amplitude. By comparing these S11 stress response maximum/minimum values with the ones shown in **Figure 28** where the response spectral curves of either the 1996-Konitsa record or the Euro-Code were employed (Section 7.1.), it can be seen that the limit (max/min) S11 stress maximum/minimum amplitudes is very similar, as expected, to the corresponding values obtained from the dynamic spectral analyses employing the 1996-Konitsa record spectral curves. As was discussed before, the Euro-Code design

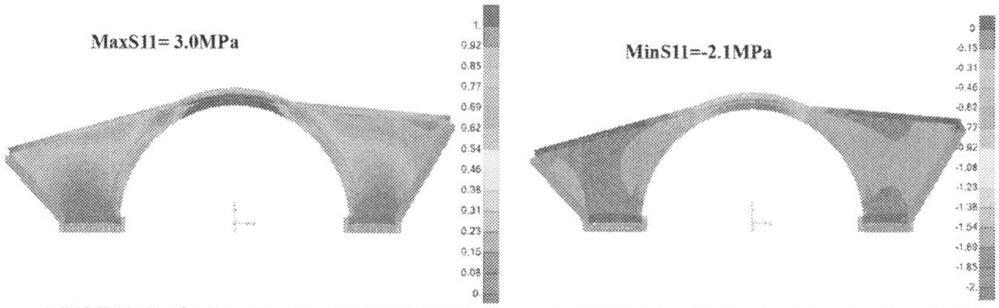

a) COMB11: Dead + Ex + Ey + Ez, MaxS11=3.0MPa
b) COMB11: Dead + Ex + Ey + Ez, MaxS11=-2.1MPa

Figure 31. State of stress through the distribution of stresses S11 (envelope) at the bottom fibre of the crown for Konitsa Bridge.

spectral curves result in much higher displacement and stress demands for the Konitsa Bridge. From all these numerical analyses, it can be concluded that the most vulnerable part of this stone-masonry bridge is the slender central part of the main arch, composed as described in Section 3 of the primary and secondary arch, when the structure is subjected to seismic forces in the horizontal out-of-plane direction. The vertical in-plane excitation is expected to be significant when in-phase with the horizontal excitation in a way that it can offset the beneficial effect of the dead weight. This observation is thought to be of a general nature, as it is demonstrated by the numerical analyses of the Plaka Bridge in the following Section 7.3.

7.3. Simplified numerical simulation of the seismic behaviour of the Plaka Bridge

This section includes results of a series of numerical simulations of the Plaka Bridge when it is subjected to a combination of actions that include the dead weight (D) combined with seismic forces. The seismic forces will be defined as was done in Section 7.1 by making use of the current definition of the seismic forces by EURO-Code 8 [12]. Towards this, horizontal and vertical design spectral curves are derived based on the horizontal design ground acceleration. This value, as it is defined by the zoning map of the current Seismic Code of Greece, is equal to 0.24 g (g is the acceleration of gravity) for the location of the Plaka Bridge [13, 14]. Furthermore, it is assumed that the soil conditions belong to category A because of the rocky site where this bridge is founded, that the importance and foundation coefficients have values equal to one (1.0), the damping ratio is considered equal to 5% and the behaviour factor is equal to 1.5 (unreinforced masonry). The design acceleration spectral curves obtained in this way are depicted in **Figure 32a** and **b** for the horizontal and vertical direction, respectively. In **Figure 32a** and **b**, the eigen-period range of the first 12 eigen-modes is also indicated (ranging between the low and the high modal period). For the vertical response spectra, this is done for only the in-plane eigen-modes (see also **Table 11**). By comparing these design spectral acceleration curves (of **Figure 32a** and **b**) for the Plaka Bridge with the corresponding spectral curves for the Konitsa Bridge (**Figure 26a** and **b**), it becomes apparent that the former represent a more demanding seismic force level than the latter.

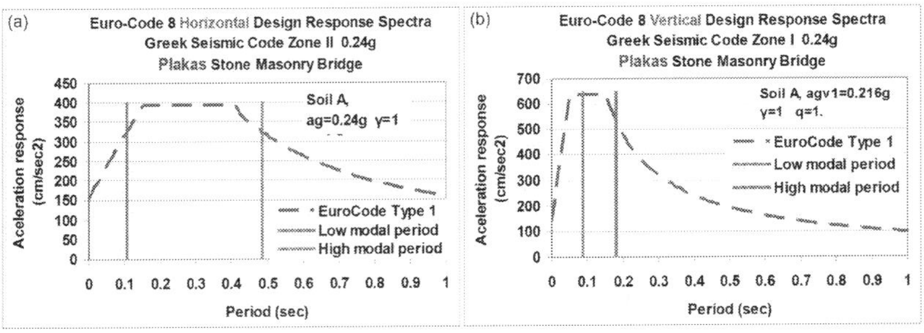

Figure 32. (a) Horizontal spectral curves for type-1 Euro-Code to be applied in Plaka bridge and (b) vertical spectral curves for type-1 Euro-Code to be applied in Plaka bridge.

For the Plaka Bridge, the modal mass participation ratios and the base reactions are listed in **Tables 12** and **13**, respectively. The base reactions are F_X, in the *x-x* (*u*1, out-of-plane), F_Y the *y-y* (*u*2, in-plane) and F_Z in the *z-z* (*u*3, in-plane) directions (see **Figures 7d**, **25**, **29a** and **b**) Apart from the dead load (D, row 1) in rows 2–4 of **Table 13**, the base reaction values were again obtained from dynamic spectral analyses employing, as was done in Section 7.1., the Euro-Code horizontal and vertical design spectral curves of **Figure 32a** and **b**. In all these dynamic spectral analyses, the 12 eigen-modes listed in **Table 12** were again employed. As can be seen in **Table 12**, these eigen-modes have modal mass participation ratios that result in sums that are $\mathrm{Sum}U_x = 67.4\%$, $\mathrm{Sum}U_y = 58.7\%$ and $\mathrm{Sum}U_z = 39.3\%$ of the total mass for the direction of motion in the U_x, U_y and U_z axes, respectively. In the subsequent load combination 1, where the dead load is combined with the Euro-Code horizontal + vertical spectral curves, the dynamic spectral analysis results were multiplied again by an amplification factor equal to the reverse of the relevant ratio values before superimposing the dead load results. This amplification factor is equal to $1/\mathrm{Sum}U_x$ for the dynamic analyses employing the out-of-plane *x-x* horizontal eigen-modal ratio, to $1/\mathrm{Sum}U_y$ for the in-plane *y-y* horizontal eigen-modal ratios and to $1/\mathrm{Sum}U_z$ for the in-plane vertical eigen-modal ratios [11]. This becomes evident when one compares the base reaction values without and with these amplification factor values in **Table 13**.

Output Case	Period	Frequency	U_X	U_Y	U_Z	SumU_X	SumU_Y	SumU_Z
Text	s	Hz	Unitless	Unitless	Unitless	Unitless	Unitless	Unitless
Mode 1 (first OOP symmetric)	0.484	2.068	0.33622	0	0	0.33622	0	0
Mode 2 (second OOP asymmetric)	0.250	3.994	0.00003	0	0	0.33625	0	0
Mode 3 (first IP asymmetric)	0.179	5.583	0	0.10253	0	0.33625	0.1025	0.000008
Mode 4 (third OOP symmetric)	0.157	6.351	0.24674	0	0	0.58298	0.10253	0.000008
Mode 5 (second IP symmetric)	0.152	6.573	0	0.00067	0.13444	0.58298	0.1032	0.13445
Mode 6 (third IP symmetric)	0.111	9.034	0	0	0.16089	0.58298	0.10323	0.29533
Mode 7 (fourth OOP asymmetric)	0.104	9.463	0.00006	0	0	0.58304	0.10323	0.29533
Mode 8 (fourth IP asymmetric)	0.0865	11.561	0	0.37475	0.00011	0.58304	0.47797	0.29545
Mode 10 (fifth IP symmetric)	0.0787	12.706	0	0.10353	0.00049	0.58304	0.5815	0.29594
Mode 9 (fifth OOP symmetric)	0.0772	12.953	0.0914	0	0	0.67444	0.5815	0.29594
Mode 11 (sixth IP asymmetric)	0.0687	14.556	0	0.00557	0.09697	0.67444	0.58708	0.39291
Mode 12 (sixth OOP asymmetric)	0.0609	16.420	0.00186	0	0	0.6763	0.58708	0.39291

Table 12. Modal participating mass ratios for Plaka Bridge (see **Figure 25**).

	Loading case description	Loading type description	Type limit	Global F_X (kN)	Global F_Y (kN)	Global F_Z (kN)
1	DEAD (D)	Linear Static		0	0	42,544
2	Euro-Code RS Hor *u*1 OP	Linear Resp. Spectral Euro-Code	Max	6382	0	0
3	Euro-Code RS Hor *u*2 IP	Linear Resp. Spectral Euro-Code	Max	0	5773	311
4	Euro-Code RS Ver *u*3 IP	Linear Resp. Spectral Euro-Code	Max	0	677	6581
5	Combination 1	Dead + Euro-Code RS (*u*1+ *u*2 + *u*3)	Max	9436	11,555	59,823
6	Combination 1	Dead + Euro-Code RS (*u*1+ *u*2 + *u*3)	Min	−9436	−11,555	25,265

Table 13. Base reactions, Plaka Bridge.

In **Figure 33a** and **b**, the numerically predicted deformation patterns of Plaka Bridge are depicted for load combination 1. As can be seen, this stone-masonry bridge develops under this combination of dead load and seismic forces relatively large out-of-plane displacements at the top of the main arch. As expected, the out-of-plane displacement response of the Plaka Bridge, when subjected to Euro-Code design spectra, reaches the largest value at the crown of the arch with a maximum value equal to 52.84 mm. This maximum out-of-plane value for the Plaka Bridge is almost twice as large as the corresponding value predicted numerically for the Konitsa Bridge.

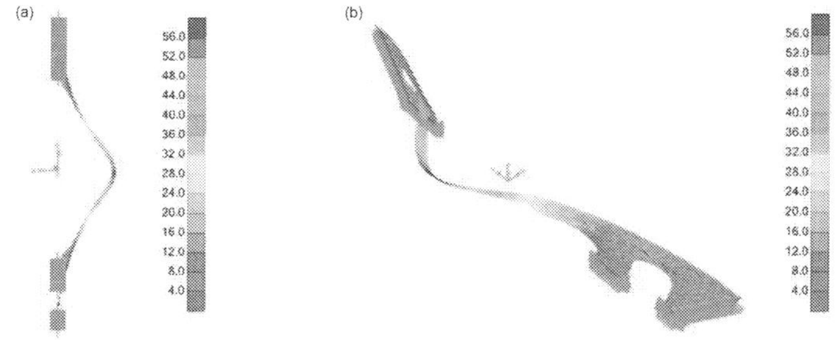

Figure 33. (a) Deformations of Plaka Bridge. For loads Dead + Euro-Code RS (*u*1+ *u*2 + *u*3). Comb 1. At crown *u*1 = −52.84 mm, *u*2 = −21.13 mm, *u*3 = −22.11 mm. (b) Deformations of Plaka Bridge. For loads Dead + Euro-Code RS (*u*1 + *u*2 + *u*3). Comb 1. At crown *u*1 = −52.84 mm, *u*2 = −21.13 mm, *u*3 = −22.11 mm.

In **Figure 34a–d**, the numerically predicted state of stress (max/min S11, max/min S22), which develops at Plaka Bridge for load combination 1, is depicted. Again, as expected, the most demanding state of stress results is for the load combination 1 that includes seismic forces provided by Euro-Code. The largest value of tensile stress S11 (5.73 MPa, **Figure 34a**) develops at the bottom fibre of the crown of the arch. This relatively large tensile stress value [11, 16] is exceeding by far the tensile capacity of traditionally built stone masonry. The largest value of tensile stress S22 (3.40 MPa, **Figure 34c**) develops at the area where the toes of the primary arch join the foundation block. Again, this is a relatively large tensile stress value and is exceeding by far the tensile capacity of traditionally built stone masonry. Both these remarks indicate locations of distress for the Plaka stone-masonry bridge, as was done for the Konitsa Bridge predicting in this way the appearance of structural damage. On the contrary, the largest value of compressive stress equal to S11 = −6.14 MPa (**Figure 34d**, for combination 1) could be met by the compression capacity of the stone masonry for this bridge. The maximum tensile stress values that were numerically predicted for Plaka Bridge are approximately twice as large as the corresponding values obtained for Konitsa Bridge. This is due to the seismic forcing levels, which for Plaka Bridge are by 50% higher than those applied for Konitsa ridge. This is because Plaka Bridge is located in seismic zone II (design ground acceleration equal to 0.24 g) whereas Konitsa Bridge is located at seismic zone I (design ground acceleration equal to 0.16 g). Furthermore, although the main central arches of the two bridges are very similar in geometry (with the deck of the Plaka Bridge being somewhat wider than the deck of the Konitsa Bridge), the Plaka Bridge has a much larger total length than the Konitsa Bridge due to the construction of a mid-pier and arches adjacent to the main central arch. Thus, Plaka Bridge is more flexible and has a much larger total mass than the Konitsa Bridge. Based on these remarks, it is reasonable to expect for the Plaka Bridge larger seismic displacement values in the out-of-plane direction and consequently larger tensile stress values, than the corresponding values predicted for the Konitsa Bridge. The final consequence of these remarks is that, according to the

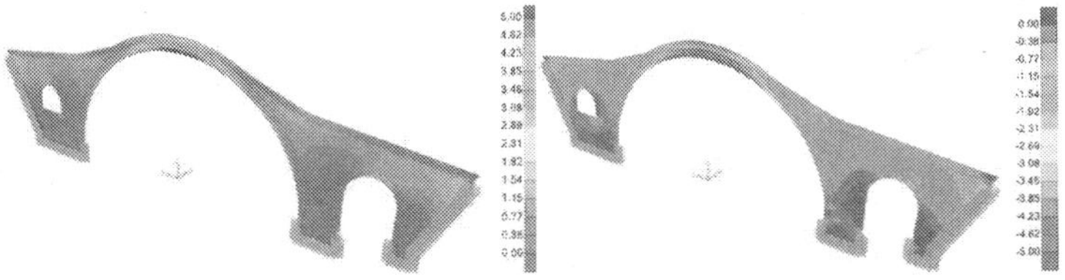

a) For loads Dead + EuroCode RS (u1+ u2 + u3)
Comb 1. max S11= 5.73MPa (Envelope Top face / Bottom face)

b) For loads Dead + EuroCode RS (u1+ u2 + u3)
Comb1. min S11= -6.14MPa (Envelope Top face / Bottom face)

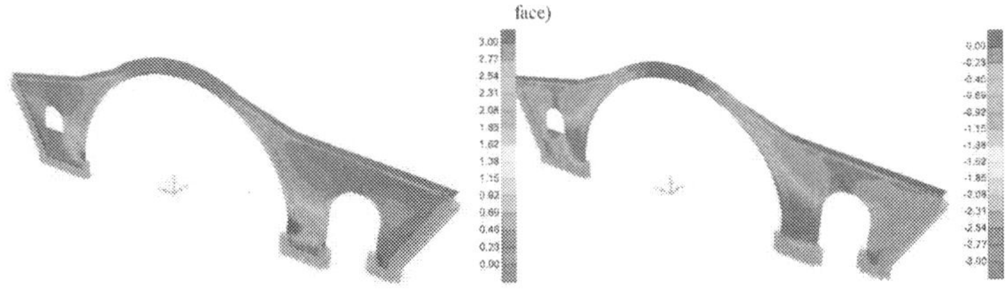

a) For loads Dead + EuroCode RS (u1+ u2 + u3)
Comb 1. max S22 = 3.40MPa (Envelope Top face / Bottom face)

a) For loads Dead + EuroCode RS (u1+ u2 + u3)
Comb 1. min S11 = -4.9MPa (Envelope Top face / Bottom face)

Figure 34. State of stress through the distribution of stresses S11 and S22 for Plaka Bridge.

results of this simplified numerical approach, the Plaka Bridge has a higher degree of seismic vulnerability than the Konitsa Bridge. A similar simplified numerical study of the performance of the Plaka Bridge could be done when measurements of flow data of the flooding of river Arachthos (31st January 2015) that caused the collapse of this bridge become available.

8. Non-linear numerical simulation of the seismic behaviour of the Konitsa stone-masonry bridge

A three-dimensional finite element model of the Konitsa stone bridge was developed and utilized in the linear (modal and gravity) and non-linear (earthquake) analyses [18, 19]. The general finite element software True-Grid (meshing) and LS-DYNA (static, modal, earthquake analyses) software were employed [20]. The developed three-dimensional model incorporated interface conditions between distinct parts of the structure (i.e. lower and upper stone arches, arches and abutments, etc.) in an effort to capture the interaction between the structural sections as well as differentiate between the building techniques and details that were introduced during the construction of the bridge and thus differentiate between the different failure criteria and mechanisms that may govern the different parts. A modelling approach where the elements (stones) of the arch are represented by solid elements with 'hybrid' behaviour was adopted and used throughout. The detailed model developed for this study included 4009 beam elements that formed the steel mesh in the intrados of the bridge rigidly connected to the stone array. The bridge was modelled using four different solid materials with 72,540 elements. As noted above, the different structural components are in 'contact' governed by contact interface conditions. The two arches have been modelled with solid 'hybrid' elements (stone-mortar behaviour) that capture the 'non-linearity' or failure rather than the pure contact between stones, an approach that is closer to the actual conditions in the structure. Specifically, it has been assumed that the 'hybrid' element representing mortar and stone behaves as one with the weakness attributed to the mortar part (Modulus and Poisson ratio represent the entire element but critical stresses are dependent on mortar).

The model is assumed to be fixed on competent rock on both sides and no soil-structure interaction (SSI) effects are considered. **Figure 35a** and **b** depict the finite element model that was developed and utilized based on *in situ* technical information collection, images and other historically available technical data. In developing the finite element method (FEM) model, special attention to the foundation and abutment details was paid and incorporated. Based on experience and data for similar structures, the first attempt in establishing the static and dynamic (modal) behaviour of the Konitsa Bridge utilized isotropic material properties for the mortar-stone material with Young's modulus E = 17 GPa, compressive strength of 30 MPa, Poisson's ratio of 0.21 and density of 2.69 g/cc. Orthotropic elastic behaviour of the hybrid stone-mortar material was also utilized in the numerical modal analysis during the calibration phase and following the field vibration test. This is described in [20] as one of the options for elastic materials but with orthotropic behaviour. **Figure 36a** depicts modelling details of the foundation of the Konitsa Bridge and of the way the primary and secondary arches are joined with the foundation block. **Figure 36b** depicts the modelling detail of the parapet and the deck of the Konitsa Bridge (see also **Figure 20**).

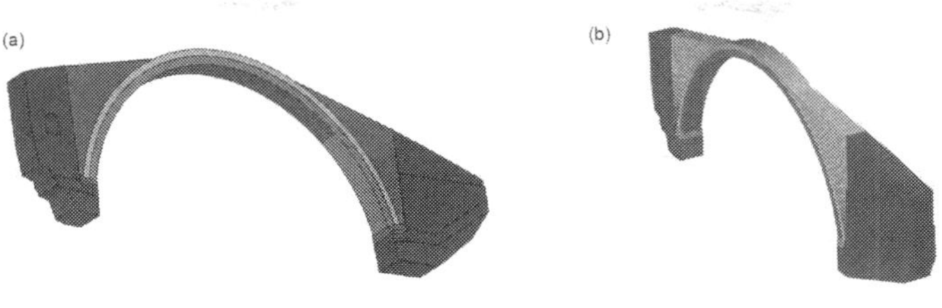

Figure 35. (a) Depiction of sections considered in modelling: the Konitsa stone arch bridge including the partial steel mesh over the intrados placed during restoration work prior to 1996 earthquake and (b) finite element model and details of Konitsa stone bridge.

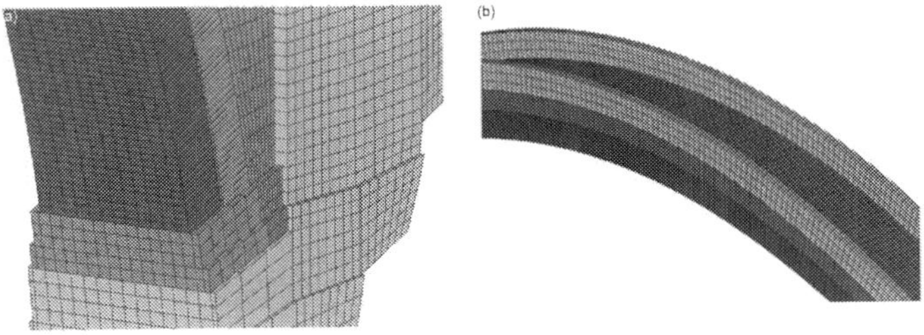

Figure 36. (a) Konitsa Bridge RHS foundation modelling details and (b) Konitsa Bridge parapet modelling details.

8.1. Modal and static analyses

Before proceeding to the complex non-linear analyses, a modal analysis was performed as a first attempt utilizing the numerical model depicted in **Figure 36a** and **b**. This was done using isotropic elastic material behaviour throughout the numerical model with material properties ρ = 2.69 g/cc, E = 17 GPa, ν = 0.21 for the two arches and similar values for the abutment and mandrel walls. The same process was followed, described in the numerical simulation of Section 6, whereby the measured eigen-frequencies, reported for this bridge in Section 4, were taken into account in the best possible way. This modal analysis led to mode and corresponding frequencies shown in **Figure 37**. The first five (5) modes include the first two bending modes, the first torsional mode, the first asymmetric vertical mode and the first pure vertical mode, as were also reported in Section 6. In what follows is again a comparison of the modal characteristics of the current 3-D numerical simulation with the results of the 3-D numerical simulation of Section 6 as well as with measured values. As can be seen from this comparison, the values of the eigen-frequencies for the out-of-plane eigen-modes compare well with the measured values, as was also discussed in Section 3. Moreover, as was also discussed in Section 6, certain discrepancies can be seen for the in-plane eigen-modes. It is believed that the use of orthotropic properties for the materials employed in both the linear numerical simulations can correct up to a point these discrepancies.

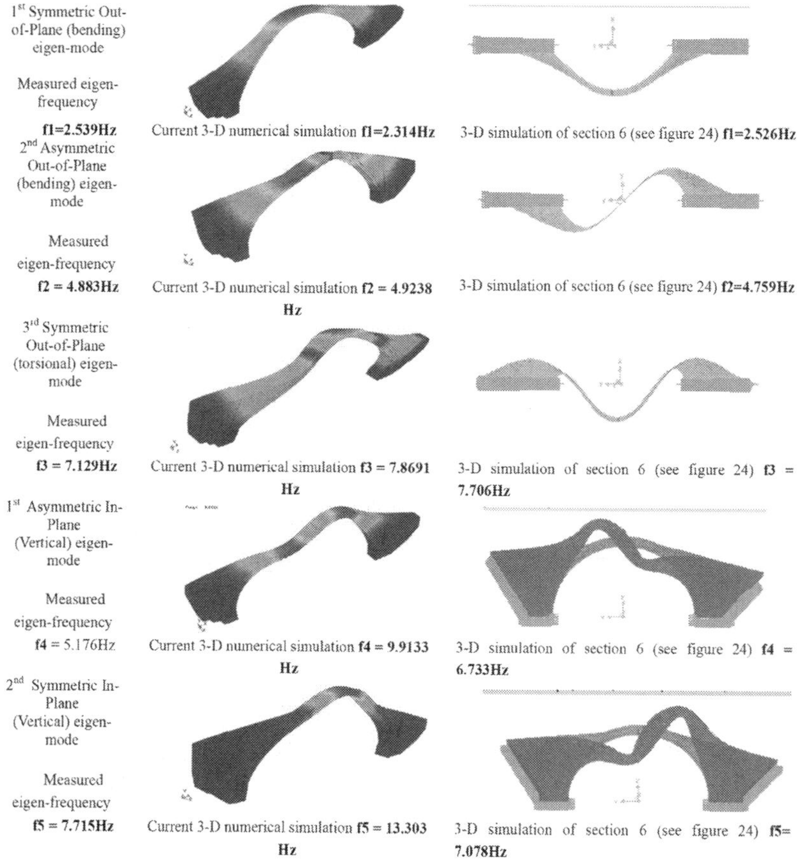

Figure 37. Comparison between numerically predicted eigen-frequencies with measured values.

8.2. Non-linear earthquake analysis and damage criteria

For the static analysis and subsequently dynamic (earthquake) analyses where the bridge structure is expected to exhibit non-linear behaviour and damage, the following material behaviour was adopted in this study.

The mortar-stone material was assumed to behave like 'pseudo-concrete' according to the Winfrith model. It is controlled by compressive and tensile strength as well as fracture energy and aggregate size.

http://dx.doi.org/10.5772/64752

The compressive strength is considered to be controlled by the stone portion of the hybrid element (30 MPa) and the tensile strength by that of the mortar. The range of the tensile strength assumed in this study for the different sections of the Konitsa Bridge is 0.25–2.1 MPa. The fracture energy assumed in the analysis dissipated in the opening of a tension crack assumed as 80 N/m. Upon formation of a tension crack, no tensile load can be transferred across the crack faces.

An additional failure criterion that controls the detachment of elements from the structure is that of pressure (negative in tension). This criterion is used to simulate the failure of mortar in the hybrid element, which is considered to fail when the negative pressure exceeds a critical value. The pressure threshold assumed in the study was 1.1 MPa.

8.3. Seismic analysis of the Konitsa Bridge

The most recent earthquake in the proximity of the Konitsa Bridge occurred in August 1996 [21]. The epicentre of the 6th August earthquake ($M = 5.7$) with 8-km depth was about 15 km to the South West (SW) of the bridge. While no recording at the bridge location is available, the earthquake was recorded at less than a kilometre away on soft soil with maximum acceleration of 0.39 g [22]. A similar recording on rock (~1 km away and on the same rock formation to that supporting the left-hand side (LHS) buttress of the Konitsa Bridge) indicated a peak ground acceleration of 0.19 g. During the 1996 earthquake, limited damage was experienced by the bridge in the form of (a) spalling of the protective cement layer in the bridge intrados that was introduced following upgrades performed a few years earlier accompanied by the introduction of a steel mesh in the intrados and (b) loss of parapet sections. A considerable number (16%) of the checked 925 buildings of the town of Konitsa, located at close proximity to the stone masonry bridge, developed structural damage typical to Greek construction [24]. The recorded ground motion (see **Figure 29** of strong motion acceleration [15]) exhibits the characteristics of an impulse-type or near-field earthquake especially its horizontal component that contains the characteristic pulse. This acceleration record, shown in **Figure 29**, is used as bridge base excitation in the non-linear analysis. Three-dimensional excitation was considered for all the seismic analyses performed. For the Konitsa 1996 earthquake analysis, the in-plane and out-of-plane horizontal components were identical and reflected the recorded horizontal acceleration trace of **Figure 29**. The vertical excitation component was the one also shown in **Figure 29**. No SSI considerations were introduced at the bottom of the two abutments, which were assumed to be fixed on rock. Further, for these analyses no differentiation in ground motion between abutment supports was considered despite the fact that one abutment is supported on competent rock and the other in what appears to be weathered rock.

The seismic study was conducted in two steps. Specifically, during the first step, the static conditions of the structure were reached by introducing a fictitiously high global damping. Upon stabilization throughout the structure (see yellow arrow in **Figure 38a**), the earthquake analysis was initiated with the correct damping estimated based on the experimental measurements made during the two campaigns (i.e. global damping of 1.6%, **Figure 38b**). **Figure 39** depicts the state-of-stress profile throughout the Konitsa Bridge due to gravity load

(**Figure 39a** depicts principal deviatoric stress, **39b** vertical stress around the right-hand side (RHS) abutment and **38a** vertical stress evolution during the gravity load analysis reaching stabilization for the start of earthquake analysis).

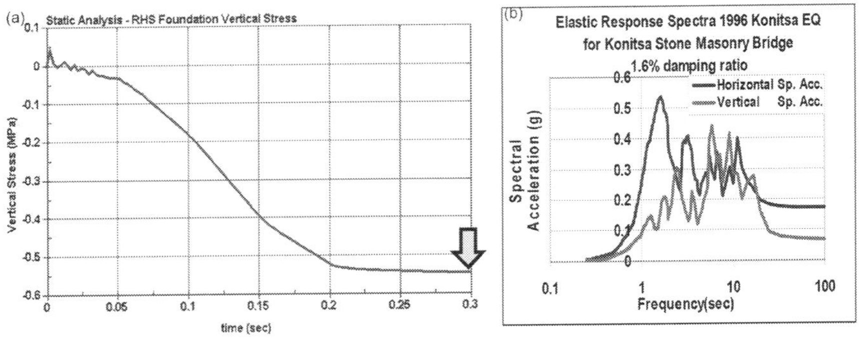

Figure 38. (a) Static state of stresses of Konitsa Bridge at the start of seismic analysis. The arrow indicates the start of the dynamic (earthquake analysis) following the gravity load analysis stabilization. (b) Two percent response spectra of the 1996-Konitsa earthquake recorded on rock.

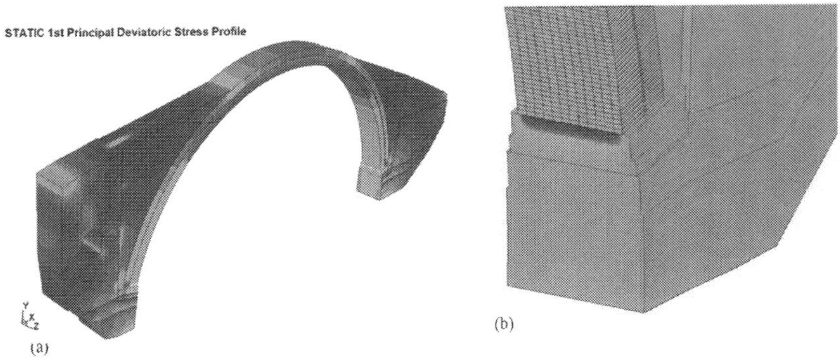

Figure 39. (a) Principal stress profile of Konitsa Bridge under dead load and (b) compressive stress concentration at the foot of the main arch.

Shown in **Figure 40a** is the location of the numerical model of Konitsa Bridge where the seismic response is predicted (crown, Loc-3, Loc-2, Loc-1) having as input motion the described seismic excitation throughout all the base points (Base EQ input). **Figure 40b** depicts the horizontal (in-plane and out-of-plane) and vertical crown displacement seismic response of the Konitsa Bridge predicted using the non-linear numerical analysis. As can be seen in **Figure 40b**, the

http://dx.doi.org/10.5772/64752

maximum predicted out-of-plane horizontal crown displacement is somewhat larger than the maximum value predicted by the linear time-history analysis in Section 7.1 (**Figure 30a**). The maximum predicted in-plane vertical crown displacement (**Figure 40b**) predicted by this non-linear earthquake analysis is significantly larger (approximately four times) than the maximum value predicted by the linear time-history analysis in Section 7.1 (**Figure 30a**). This must be attributed to the fact that the linear analysis performed in Section 7.1 is three-dimensional but employing a numerical model of the bridge that represents its mid-surface, whereas the 3-D non-linear simulation utilizes a model where the bridge is simulated with its actual thickness (compare **Figure 15** with **Figures 35** and **36**). Thus, the vertical displacement at the crown (see **Figure 40a**) predicted by the 3-D non-linear analysis represents the vertical displacement at the façade of the crown cross section of the bridge, which includes a contribution from the out-of-plane response, and not the vertical displacement of the crown at mid-surface, as is the case for the simplified analysis of Section 7.1 (**Figure 30a**). The in-plane horizontal displacement predicted by both the linear and the non-linear earthquake analyses has relatively very small amplitude. As discussed before, this clearly demonstrates the much larger stiffness of the bridge structure along the horizontal in-plane direction than along the out-of-plane direction. In **Figure 41a** and **b**, the absolute velocity response at four locations of the Konitsa Bridge as well as at its base is depicted in the horizontal in-plane or out-of-plane direction, respectively. As can be seen again in these figures, the stiffness of the bridge combined with the applied seismic motion results in very small amplification of this velocity response in the in-plane direction than in the out-of-plane direction between the base and the four Konitsa Bridge locations (Crown, Loc-3, Loc-2, Loc1). This crown/base velocity response amplification factor in the out-of-plane direction has a value approximately equal to 2.

In **Figure 43**, the contours of the effective von-Mises stresses are depicted for the Konitsa Bridge subjected to the previously described 1996-Konitsa earthquake record. As can be seen in this figure, tensile distress is indicated at the right and left ends of the primary and secondary arches where they join the foundation blocks. This is also shown in some detail in **Figure 42a** and **b** in terms of von-Mises and vertical stress response in this location. The time-history plot of the vertical stress at the foundation block (A) at the arch-to-foundation block interface (B) and at the primary arch (C) clearly indicates that the tensile stress at location C reaches, as expected, the largest value, which is in excess of the tensile capacity of the bridge construction material (see also Section 7.1. and **Figure 28e**). By comparing the results of the displacement and stress response of the Konitsa Bridge, as obtained by the present 3-D non-linear analysis, with the corresponding time-history results of the simplified linear analysis of Section 7.1, it can be concluded that the 1996-Konitsa ground motion employed in both cases was of such an intensity and frequency content that very limited non-linearities developed at this 3-D advanced non-linear model of the structure. This conclusion is in line with the observed performance of this bridge during the 1996 main event. As already mentioned before, limited damage was experienced by this bridge in this 1996 earthquake in the form of (a) spalling of the protective cement layer in the bridge intrados that was introduced following upgrades performed few years earlier and (b) loss of parapet sections (**Figure 43**).

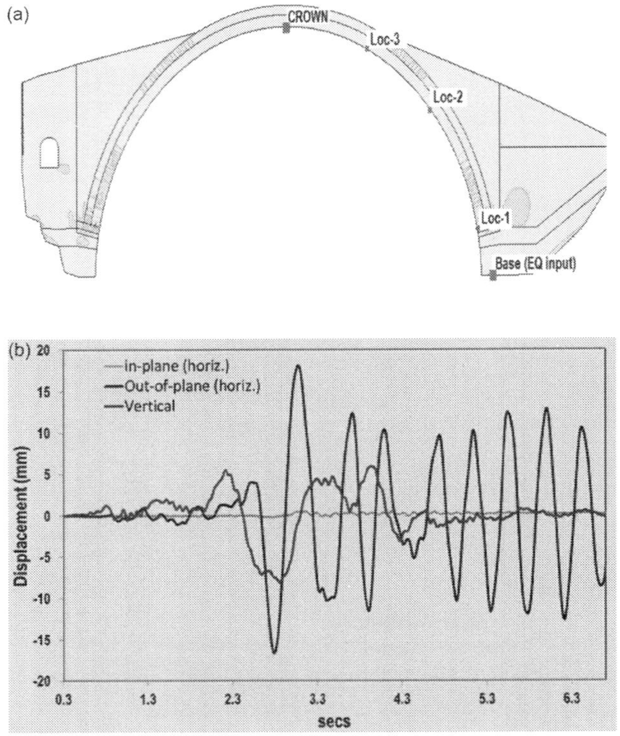

Figure 40. (a) 3-D model of Konitsa Bridge together with the locations of input (excitation) and predicted seismic response and (b) horizontal (in-plane and out-of-plane) and vertical crown displacement seismic response of the Konitsa Bridge predicted using the non-linear numerical analysis.

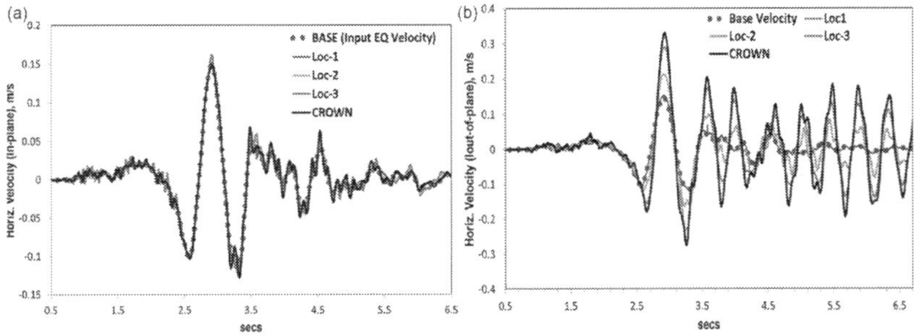

Figure 41. Earthquake response of Konitsa Bridge when subjected to the 3-D 1996-Konitsa earthquake: (a) in-plane horizontal velocities and (b) out-of-plane horizontal velocities.

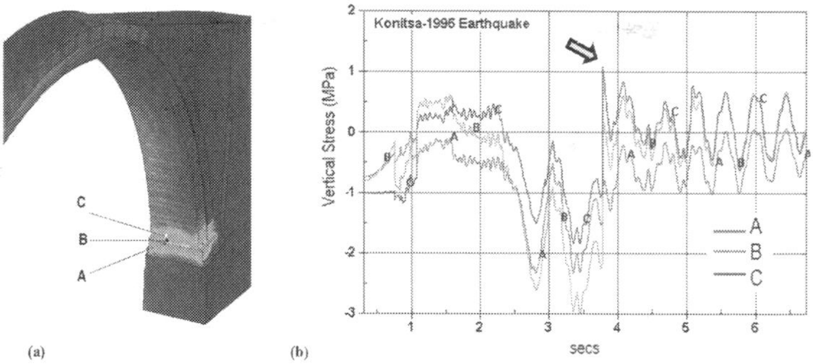

Figure 42. Vertical stress response at the right end of the primary and secondary arches where they join the foundation blocks.

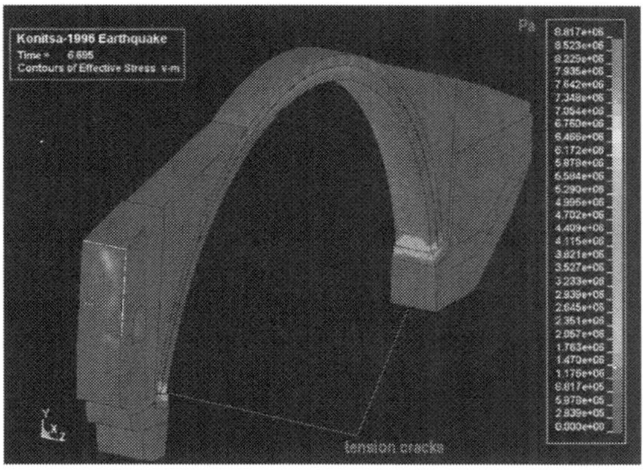

Figure 43. Tensile stress concentration at the foot of the primary and secondary arches of the Konitsa Bridge.

8.4. Seismic vulnerability assessment and code guidance effects

In order to examine the capabilities of the 3-D non-linear numerical simulation performed in the previous section and in an effort to understand the potential influence of the time structure and period content of the exciting earthquake which may be missed when utilizing envelope code spectra (i.e. Euro-Code), the Konitsa Bridge was subjected to two (2) additional earthquakes that represent distinct classes, namely near-field (impulsive-type) and far-field earthquakes. Specifically, the NS component observed at Shiofukizaki site in the 1989 Ito-Oki earthquake of moment earthquake magnitude 5.3, epicentral distance of 3 km and the depth of the seismic source of 5 km. The record was observed at the surface of basalt rock and has a maximum acceleration of 0.189 g. It has been characterized as a near-field earthquake and it exhibits remarkable similarity to the Konitsa 1996 earthquake (**Figure 44**, top).

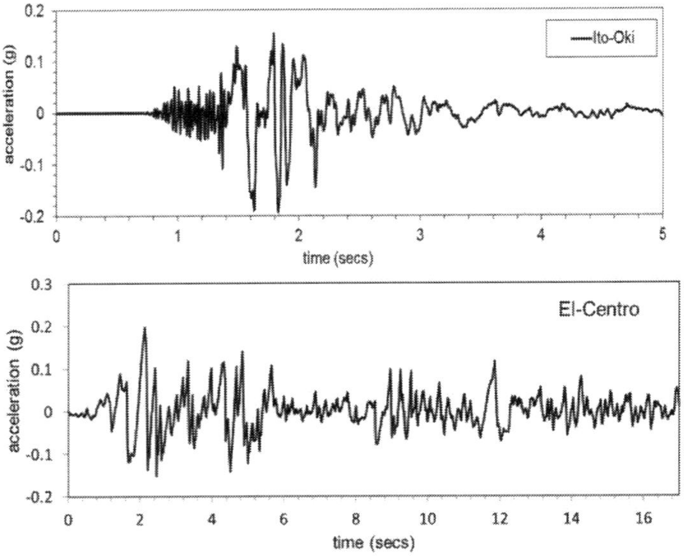

Figure 44. Acceleration time histories of Ito-Oki 1989 and El-Centro1940 PGA-adjusted earthquakes.

The second earthquake is the 1940 El-Centro normalized to 0.19 g (**Figure 44**, bottom) allowing for direct comparison with the similar PGA Konitsa-1996 and Ito-Oki near-field earthquakes. The direct comparison of the response spectra of the three earthquakes (Konitsa-1996, Ito-Oki and normalized 1940 El-Centro) is shown in **Figure 45**. The objective of subjecting the Konitsa Bridge to the same PGA but different spectral content earthquakes is to directly compare the damageability potential based on the non-linear response of the bridge and shed some light on sensitivities to the type of earthquake these type of structures (masonry stone bridges) exhibit. This ultimately will aid in the modification/updating of the seismic codes to capture the unique structural design and response characteristics of large span arch masonry bridges in their provisions. While for the Konitsa-1996 earthquake the actual vertical acceleration was used, for the Ito-Oki and modified 1940 El-Centro the vertical component was assumed as 75% of the employed horizontal component. The results drawn from the three (3) non-linear analyses (Konitsa-1996, Ito-Oki and 1940 El-Centro) and the comparative damageability potential are very revealing. Specifically, very similar response and bridge damage are observed for the two impulsive-type earthquakes, M =5.7 Konitsa-1996 and M = 5.3 Ito-Oki earthquakes, which are similar PGA and time structure. Their damage potential is quite limited and it confirms the observations made post 1996-Konitsa earthquake of the bridge. **Figure 46a** and **b** depict the Konitsa Bridge out-of-plane displacement and stress response, respectively. On this basis and by comparing these maximum response values with the corresponding maximum values obtained utilizing the 1996-Konitsa earthquake record as input motion (**Figures 30a** and **b**, **40b**, **41a** and **b**), it can be concluded that the potential damage vulnerability from the Ito-Oki earthquake resembles that of the Konitsa-1996 earthquake.

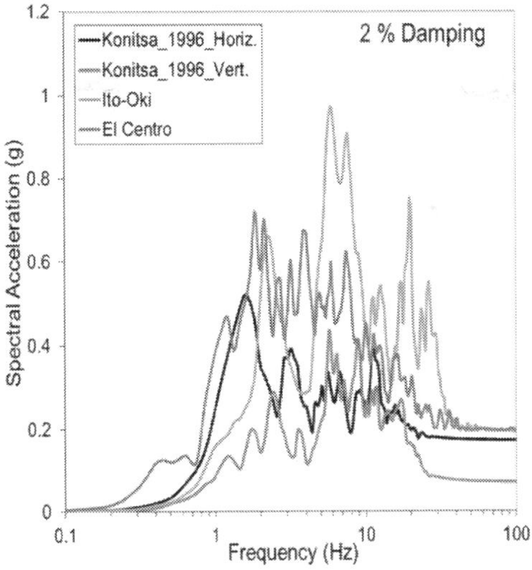

Figure 45. Acceleration response spectra (2% damping) of the similar PGA but different type (near- vs. far-field) earthquakes utilized in the study.

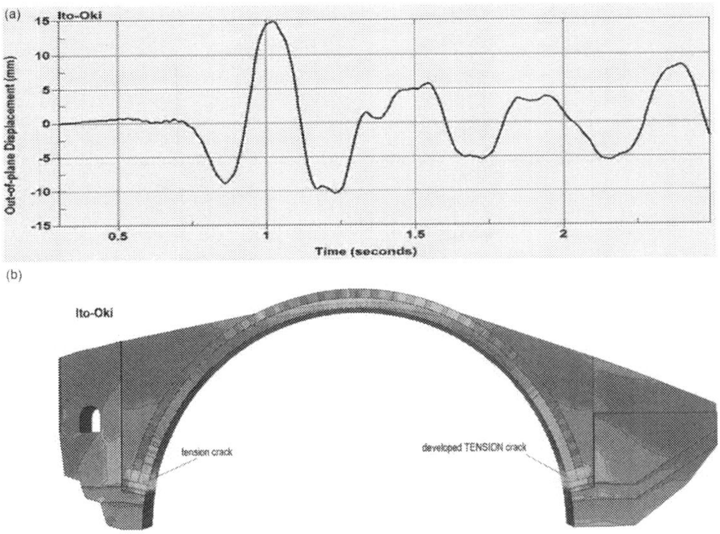

Figure 46. Konitsa Bridge response to the M = 5.3 Ito-Oki (0.19g PGA) near-field (impulsive) earthquake. (a) Out-of-plane displacement response and (b) tensile stress response.

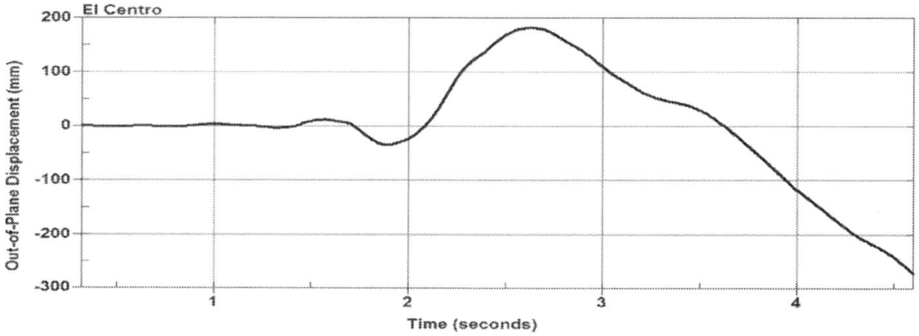

Figure 47. Konitsa Bridge out-of-plane displacement response to PGA-adjusted (0.19 g) 1940 El-Centro earthquake.

A strikingly different bridge response and damage potential are observed when Konitsa Bridge is subjected to an excitation with the 0.19-g normalized 1940 El-Centro earthquake, which represents a different type (far-field) of seismic event lacking that characteristic dominant velocity pulse (**Figure 29**). **Figures 47** and **48** clearly demonstrate the different damage potential of this type of earthquake on such relatively long-span stone-masonry bridges. **Figure 47** depicts the Konitsa Bridge out-of-plane displacement response when subjected to PGA-adjusted (0.19-g) 1940 El-Centro earthquake. **Figure 48** depicts the variation of tensile stresses together with relevant non-linear deformations of large amplitude at critical locations of the main arch during certain time windows of the response when these deformations are maxi-

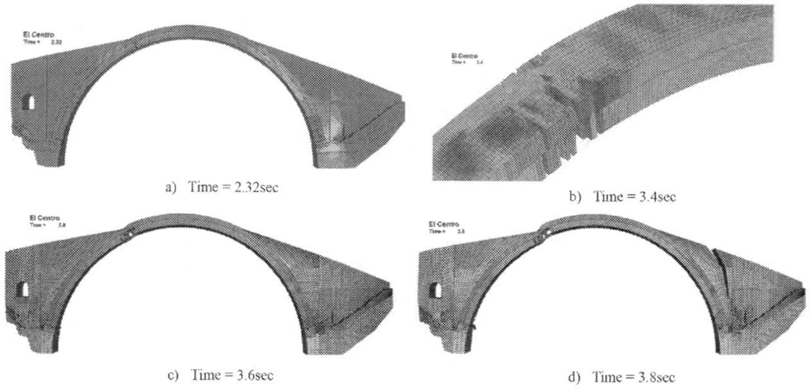

Figure 48. Evolution of damage resulting from the El-Centro (0.19-g) far-field-type earthquake. (a) Time = 2.32 s, (b) 3.4 s, (c) 3.6 s and (d) 3.8 s.

mized, thus indicating the collapse potential of that portion of the main arch. This prediction of the main arch performance is in agreement with a similar conclusion reached by the simplified analyses of Section 7.1 when Konitsa Bridge (**Figures 27b**, **28b** and **d**) and Plaka Bridge (**Figures 33** and **34**) were subjected to the design spectra as defined employing provisions of Euro-Code 8. This large variation in the damageability potential, therefore, should be accounted for in establishing seismic code guidelines for relatively fragile old cultural heritage structures (as the old stone-masonry bridges studied in this chapter) as they apply to these non-typical structures. It should be noted that similar conclusions regarding the damageability of near-field-type earthquakes, as compared to their far-field counterparts based on which seismic codes for nuclear structures were deduced, were reached following an International Atomic Energy Agency (IAEA)-launched coordinated research project (CRP) experimental study [23] augmented with numerical analysis and response/damage predictions conducted by an international participation.

9. Maintenance issues for stone-masonry bridges

In this section, a brief discussion will be presented dealing with maintenance issues of the stone-masonry bridges that were examined in this chapter. This study focused on the dynamic and seismic response of this type of bridges. However, it was shown by past experience that structural damage can also result from other types of actions such as flooding or traffic when such bridges are used not only for light pedestrian use. Because almost all the stone-masonry bridges in Greece have been built mostly for relatively light live load levels resulting from the crossing of pedestrians or animal flocks, their structural vulnerability due to traffic conditions is not an issue. Instead, flooding of the narrow gorge currents that these bridges cross (**Figure 49a**) is one of the main structural damage causes, as demonstrated from the Plaka Bridge (see **Figure 50a** and **b**). Apart from the hydrodynamic loads that a stone-masonry bridge is subjected to from a flooded current, one of the main sources of distress that may lead to partial or total collapse is the deformability of the foundation. The deformability of the foundation and the potential for subsequent collapse does include not only wash-out effects from a sudden flooded current but also the cumulative deformability of the foundation in a wider time

Figure 49. (a) Almost total flooding of a stone masonry bridge and (b) tilting of a mid-pier and partial collapse of the Diava-Kalampaka reinforced concrete bridge in Thessaly, Greece (16th January, 2016).

window as was demonstrated by a recent flooding of Pineios river that caused the tilting of a mid-pier and the partial collapse of the Diava-Kalampaka-reinforced concrete bridge in Thessaly, Greece (16 January 2016, **Figure 49b**).

Figure 50. (a) View of the Plaka Bridge after the collapse from the West bank. Note the total destruction of the mid-pier (see also **Figure 7d**). (b) Close-up of the total destruction of the mid-pier of Plaka Bridge (see also **Figure 7d**).

Thus, foundation maintenance seems to be of the utmost importance. The flooding of Arachthos river, which caused the collapse of Plaka Bridge on 31 January 2015, was of considerable proportions. It is of interest to observe the conditions of the mid-pier of Plaka Bridge after the collapse (**Figure 50a** and **b**). As can be seen, the foundation of this pier is almost non-existent being covered by the remains of the East part of central arch and of part of the adjacent arch and mid-pier. Thus, it is evident that this mid-pier was highly distressed leading to this mode of collapse.

Another maintenance issue of considerable importance is the integrity of the stone masonry in parts of the bridge apart from the foundation. It was already discussed in Section 6, when comparing numerically predicted with measured eigen-frequency values, that evidence of washed-out mortar joints was present mainly in Kontodimou, Tsipianis and Kokorou Bridges. At the time of *in situ* measurements (October to December 2015), maintenance works took place in Konitsa Bridge focusing on the removal of vegetation and re-pointing of the mortar joints. The effectiveness of these operations must be validated through laboratory testing regarding the compatibility and durability of the materials employed. The presence of metal ties and their structural function was underlined in Section 3. However, inspection of these metal ties in the stone-masonry bridges of the present study as well as other stone-masonry bridges not reported here casts doubts on their effectiveness due to lack of maintenance for a long time.

In some cases, these stone-masonry bridges suffered structural damage from human activity. Plaka Bridge is one such example as can be seen in **Figure 51a**. The red arrow in this figure points to the structural damage suffered by the central arch due to an explosion during World War II. The damaged part was retrofitted in a way that is not known in detail to the authors. This retrofitting is visible in detail in **Figure 51b** where one can distinguish the difference in the texture of the old stone masonry from the retrofitted part of the secondary arch in this location indicated by the red circle.

http://dx.doi.org/10.5772/64752

Figure 51. (a) Structural damage at the East part of the central arch of Plaka Bridge due to an explosion. (b) Detail of the retrofitted part of the secondary arch of the Plaka Bridge sometime before its collapse from flooding.

This is also visible in **Figure 52a** where the scaffolding used for additional maintenance works is also visible. However, these works did not prove sufficient to prevent the collapse of this bridge from the severe flooding. It is of great research interest to be able to apply the methodology of *in situ* investigation presented in Section 4 of this chapter together with a long-term monitoring and maintenance programme as means of safeguarding the structural integrity of these precious cultural heritage structures.

Figure 52. (a) Maintenance works at Plaka Bridge sometime before its collapse from flooding and (b) stone-masonry bridge at Dasilio-Grevena, Greece after being retrofitted.

In the brief space of this section, the principles that govern a major retrofitting of such bridges must also be underlined. This is very important not only for the collapsed Plaka Bridge and the plans for its reconstruction but for numerous other bridge structures that have suffered serious structural damage or partial collapse. **Figure 52b** depicts a stone-masonry bridge in North-Western Macedonia, Greece, which underwent major reconstruction. It is worth mentioning that the regions of North-Western Macedonia and Ipiros in Greece are the home of stone masons who have been active worldwide. Due to their initiative specific stone-masonry workshops have been established recently in this region in an effort to keep this type of traditional construction as well as its maintenance alive.

10. Conclusions

This structural performance of old stone-masonry bridges is studied following a methodology that utilizes a large number of high fidelity *in situ* measurements in order to identify their dynamic characteristics (eigen-frequencies, eigen-modes and damping ratio) and corresponding numerical predictions from relatively simple or more complex numerical simulations. The validity of these numerical simulations was ascertained by comparing the measured *in situ* dynamic response with the one predicted numerically. This methodology was extended by applying such 'realistic' numerical simulations to predict the performance of specific old stone-masonry bridge structures (e.g. Konitsa Bridge) when subjected to dead load combined with seismic actions. A series of numerical dynamic analyses, both simplified (linear) and complex (non-linear), were made. In these analyses, actual earthquake excitation recorded in the proximity of Konitsa Bridge and relevant seismic code design seismic spectra were employed as well as earthquake records representing near-field (impulsive-type) or far-field seismic events. Seismic actions specified in these ways were used to investigate the damage potential of such stone-masonry bridges. It is believed that it is of great research interest to be able to apply this methodology together with measurements from a long-term monitoring and maintenance programme as a means of safeguarding the structural integrity of these precious cultural heritage structures. Finally, recommendations for intervention works should include clauses providing for preparatory actions of measurements and analyses similar to the methodology presented here. The same methodology can be applied to address flooding, which is also one of the main causes of structural damage for stone-masonry bridges that require special attention and is the subject of a separate study. Apart from the hydrodynamic loads that a stone-masonry bridge is subjected to from a flooded current, one of the main sources of distress that may lead to partial or total collapse is the deformability of the foundation. Finally, the integrity of the stone masonry in various parts of such old stone-masonry bridges is an additional maintenance issue of considerable importance. The following represent additional conclusive remarks:

1. The numerically predicted bridge deformation and stress state seismic response are in good agreement resulting from either the simple or complex numerical simulations as well as with observed structural performance following actual earthquake occurrence in the proximity of Konitsa Bridge. This offers confidence in the described methodology using (a) detailed modelling which incorporates both field measurements of the dynamic characteristics and (b) laboratory testing on the complex mortar-stone mechanical behaviour in its aged/weathered condition.

2. The high fidelity of the complex non-linear numerical analyses that were used to predict the vulnerability of these structures to earthquakes and account for new fault information that surfaces in the proximity of these structures should also be underlined. The influence of certain issues that were not included in the current numerical treatment, such as soil-structure interaction, deformability of the foundation, and so on, should also be addressed in the future. This vulnerability analysis demonstrated:

- The damage potential of far-field earthquakes on these types of unique structures (long-span masonry stone) is far greater than the damageability of impulsive-type earthquakes. This is confirmed through detailed analysis of the Konitsa Bridge using actual impulsive and far-field seismic records.
- The observations on the damageability variation between impulsive- and far-field earthquakes confirm previously conducted experimental and numerical studies on nuclear structures.
- The above findings should be considered in establishing seismic code guidelines to specifically apply to these structures considering that they are typically constructed to span river beds that are in turn closely related to faulting [25].

Acknowledgements

- The assistance of D. Gravas, K. Giouras and C.G. Manos junior in conducting the field experiments and gathering geometric information relevant to the stone-masonry bridges presented here as well as the help of the local people is gratefully acknowledged.
- For conducting the laboratory tests, we would like to thank T. Koukouftopoulos and V. Kourtidis from the laboratory of Strength of Materials and Structures of Aristotle University.
- Finally, we would also like to thank J. Evison Manou for editing this manuscript.

Author details

George C. Manos[1*], Nick Simos[2] and Evaggelos Kozikopoulos[1]

*Address all correspondence to: gcmanos@civil.auth.gr

1 Laboratory of Experimental Strength of Materials and Structures, Department of Civil Engineering, Aristotle University of Thessaloniki, Thessaloniki, Greece

2 Brookhaven National Laboratory, Upton, NY, USA

References

[1] O'Connor C., (1993), Roman bridges, Cambridge University Press, ISBN 0-521-39326-4.

[2] Grassos G. Editor, (2007), The stone masonry arch bridges of Greece, Center of Environmental Education, Makrinitsas, ISBN: 978-960-98043-9-4 (in Greek). http://kpemakrin.mag.sch.gr. Published by Eptalofos, www.eptalofos.gr.

[3] Psimarni K., Georgopoulos A., Balodimos D.D. (2000), "Development of a geographic information system for the traditional bridges of central Zagori", Report to the Municipality of Zagori, in Greek.

[4] Aoki T., et al, (2007), "Theoretical and experimental dynamic analysis of Rakanji Stone Arch Bridge, Honyabakei, Oita, Japan," 7th International Conference on Motion and Vibration Control, MOvIC 04.

[5] Sevim Barıs, et al., (2011), "Finite element model calibration effects on the earthquake response of masonry arch bridges," Finite Elements in Analysis and Design, 47 (2011), 621–634.

[6] G. C. Manos G.C., Pitilakis K.D., A. G. Sextos A.G., V. Kourtides V., Soulis V., and Thauampteh J., (2015), "Field Experiments for Monitoring the Dynamic Soil–Structure–Foundation Response of a Bridge-Pier Model Structure at a Test Site", J. Struct. Eng. 141(1), D4014012; http://dx.doi.org/10.1061/(ASCE)ST.1943-541X.0001154.

[7] Manos G.C. and Kozikopoulos E., (2015), "In-situ measured dynamic response of the Bell Tower of Agios Gerasimos in Lixouri-Kefalonia, Greece and its utilization of the numerical predictions of its earthquake response", COMPDYN 2015, Greece, 25–27 May 2015.

[8] Ozden Caglayan B., Kadir Ozakgul and Ovunc Tezer, (2012), "Assessment of a concrete arch bridge using static and dynamic load tests", Structural Engineering and Mechanics, Vol. 41, No. 1 (2012), 83–94.

[9] Simos N. and Manos G.C., "Earthquake Vulnerability of Stone Arch Bridges using Nonlinear Finite Elements and Measurements of Dynamic Characteristics," *Engineering Structures*, 2016 (submitted, in-review).

[10] Simos N. and Manos G.C, (2013), "Numerical analysis of seismic response of natural stone arch bridges-field observations and a case study," COMPDYN 2013, http://www.eccomasproceedings.org/cs2013/.

[11] Manos G.C. and Kozikopoulos E., (2015), "The dynamic and earthquake response of basilica churches in Kefalonia, Greece including soil-foundation deformability and wall detachment", COMPDYN 2015, Greece, 25–27 May 2015.

[12] Eurocode 8 – Design of structures for earthquake resistance – Part 2: Bridges, DRAFT, No. 3. European Committee for Standardization; Management Centre: rue de Stassart, 36 B-1050 Brussels, 2004.

[13] Provisions of Greek Seismic Code 2000, OASP, Athens, December 1999. Revisions of seismic zonation introduced in 2003.

[14] Paz. M., (1994), *International Handbook of Earthquake Engineering:* "Codes, Programs and Examples", edited by Mario Paz, Chapter 17, Greece by G.C. Manos, Chapman and Hall, ISBN 0-412-98211-0, 1994.

[15] Institute of Engineering Seismology and Earthquake Engineering (ITSAK), Data Base of Greek Earthquake Strong Motions, http://www.itsak.gr/en/head.

[16] Manos G.C., Kotoulas L., Soulis V., Felekidou O., (2015), "Numerical simulation of the limit non-linear behaviour of unreinforced masonry under in-plane state of stress from gravitational and seismic actions", COMPDYN 2015, Greece, 25–27 May 2015.

[17] Kiyono J., et al., (2012), "Seismic Assessment of Stone Arched Bridges," 15 WCEE, Lisbon, Portugal, 2012.

[18] Drosopoulos G.A., Stavroulakis G.E., Massalas C.V., (2006), "Limit analysis of a single span masonry bridge with unilateral frictional contact interfaces," Engineering Structures 28 (2006) 1864–1873.

[19] Korompilias D., (2015), "Study of the inelastic behaviour of the Konitsa Bridge using an inelastic model for masonry and applying strengthening methods," PhD Thesis, University of Patras, Greece, 2015 (in Greek).

[20] LS-DYNA-Version 9.71, Livermore Software Technology Corp. – LSTC, True-Grid-Version 2.3.4, XYZ Scientific Applications, Inc.

[21] Papanastasiou D., (2001), "The Konitsa, Epirus NW Greece, July 26 (Ms = 5.4) and August 5, 1996, (Ms = 5.7) earthquakes sequence", Bulletin of the Geological Society of Greece, XXXIV, 1555–1562.

[22] Spyrakos C.C., Maniatakis C.A. and Taflambas J., (2008), "Evaluation of near-source seismic records based on damage potential parameters: Case study: Greece", Soil Dynamics and Earthquake Engineering 28 (2008) 738–753.

[23] Non-linear Response to a Type of Seismic Input Motion, IAEA-TECDOC-1655-ISSN 1011-4289, June 2011.

[24] Manos G.C., (2011), "Consequences on the urban environment in Greece related to the recent intense earthquake activity", International Journal of Civil Engineering and Architecture, December, Vol. 5, No. 12 (Serial No. 49), pp. 1065–1090.

[25] Galanakis D., Paschos P., et al., (2007), "Neotectonic Activity of Konitsa Area and the 1996 Earthquakes", Hellenic Journal of Geosciences, Vol. 42, 57–64.

Chapter 5

Different Solutions for Dissipation of Seismic Energy on Multi-Span Bridges

Alessandro Contin and Andrea Mardegan

Additional information is available at the end of the chapter

http://dx.doi.org/10.5772/63821

Abstract

This chapter deals with the application of various methods for the dissipation of seismic energy in order to adjust the response to seismic forces of an existing bridge with multiple spans.

Specifically, it offers a comparison between elastomeric isolators, curved surface sliders and viscous dampers devices.

The goal is to keep the substructures and the deck substantially elastic for the limit state conditions, required by standards. Different methods will be compared by examining the interferences with traffic and with existing structures and of the worksite operational needs of supply time and of costs.

Keywords: retrofit, existing bridge, seismic energy dissipation, isolation, devices comparison

1. Introduction

Many methods have been proposed for mitigating the harmful effects of strong earthquakes. The conventional approach requires that structure passively resist through a combination of strength, deformability and energy absorption. The level of damping in these structures is typically low. During earthquakes, these structures deform well beyond the elastic limit and remain intact due to their ability to deform inelastically. The inelastic deformation takes the form of localised hinges, which results in an increased flexibility and energy dissipation. Therefore, much of the earthquake energy is absorbed by the structure through localised damage of lateral force-resisting system.

An alternate approach to mitigate the hazardous effects of earthquakes is based on a consideration of the distribution of energy within a structure. The input energy from the ground acceleration is transformed into both kinetic and potential (strain) energy that must be either absorbed or dissipated trough heat. A large portion of the input energy, instead of being absorbed by hysteretic action (i.e. damage of the structure) can be dissipated with supplemental systems.

This approach to seismic energy dissipation is made clear by considering the following time-dependent conservation of energy relationship:

$$E(t)= E_k(t) + E_z(t) + E_h(t) + E_d(t) \quad (1)$$

where E is the absolute energy input from the earthquake motion, E_k is the absolute kinetic energy, E_s is the elastic (recoverable strain energy), E_h is the irrecoverable energy dissipated by the structural system through inelastic or other forms of action (hysteretic or viscous), E_d is the energy dissipated by a supplemental damping system and t represents the time.

The chapter analyses the seismic retrofit of an existing viaduct, by assessing an actually designed case study. The viaduct at issue, built during the Seventies, serving a fast-moving thoroughfare in the city of Naples (Italy), features reinforced concrete piers and a mixed steel-concrete box deck. Three solutions, corresponding to the most common types of interventions for earthquake protection, have been compared: the first by using hydraulic dissipation devices, the second using sliding pendulum isolator devices and the third by using elastomeric isolators. The differences have been compared according to traffic interferences, interferences with existing structures, operational needs and to time and costs necessary to achieve a final project solution.

1.1. The viaduct main features

The viaduct analysed, which already has an earthquake intervention in place, features a length of 1360m and a continuous deck on 18 spans with variable clearance from 62 to 92m, separated by four longitudinal joints, implemented with Gerber saddles to assure structural continuity (**Figures 1** and **2**). The reinforced concrete piers have a hollow circular section, with a variable base diameter ranging from 2.65 to 3.30m. They are vertically compressed and feature a variable height ranging from 7.90 to 38.80m. The foundations are deep, well and raft based with piles. The structural joints are located at the piers 4, 6, 11 and 14. One particular aspect of the viaduct is that there is only one support per each pier, placed at the central part of the box. Hence, the deck is balanced, since the two adjacent runways are connected by means of transverse girders at each support section (**Figure 3**).

Intervention on an existing work always requires special attention and a meticulous study. For example, with the viaduct under examination, the base of the piers is inaccessible for any type of reinforcement intervention. Additionally, it is not possible to stop traffic in order to replace the supports, and therefore, it is not possible to work on the abutments, to draw back the ballast retainers if a larger gap than the present one (measuring 15 cm) should be required. Moreover,

http://dx.doi.org/10.5772/63821

the client wishes that the bearing system, present in the current situation, is respected, so that during all the installation phases of the new devices, the viaduct does not change its static operating conditions and above traffic can continue uninterrupted (represented in **Figure 5**).

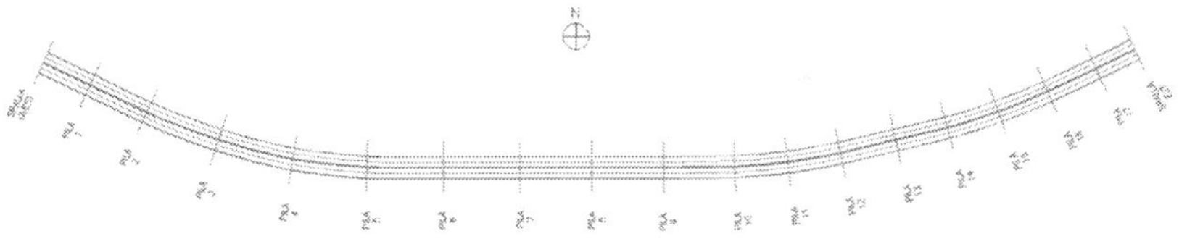

Figure 1. Viaduct plan.

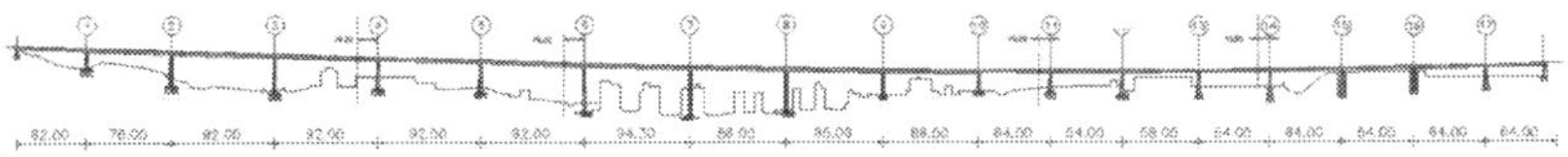

Figure 2. Viaduct view.

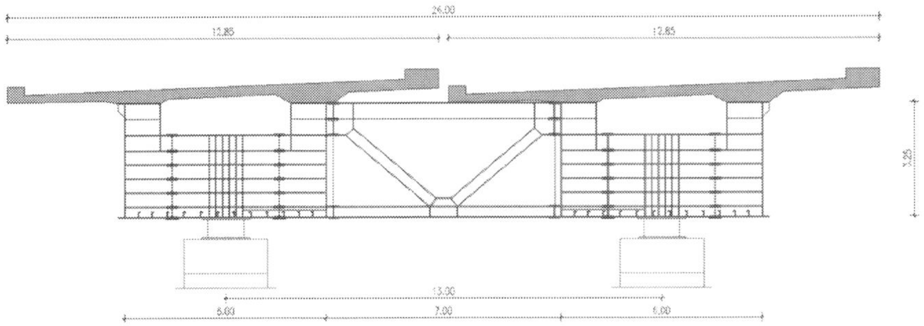

Figure 3. Typical cross-section of the viaduct.

2. Modelling

The analyses have been carried out using a finite-element analysis software on a model, consisting in mono-dimensional elements; they are non-linear with step-by-step integration of the equations of motion.

For seismic purposes, the following parameters (Italian Standard DM 14.01.2008 [1, 2]) have been taken into account:

- Rated life of the work: $V_N = 50$ years, relative to bridges and infrastructures;
- Use coefficient: $C_U = 2$, corresponding to Use Class IV about strategic construction in case of calamity and bridges of particular importance after seismic events in order to keep communication ways available;
- Reference life of the work equals to V_R=100 years; it comes from the product $V_R = V_N \cdot C_U$;
- Soil type C: deposits of medium thickened soil, with layers of more than 30m where mechanical properties gradually increase with depth; VS30 (average equivalent velocity of S wave in the first 30 m) range is 180—360 m/s;
- Topographic coefficient T_1=1 that correspond to a flat surface.

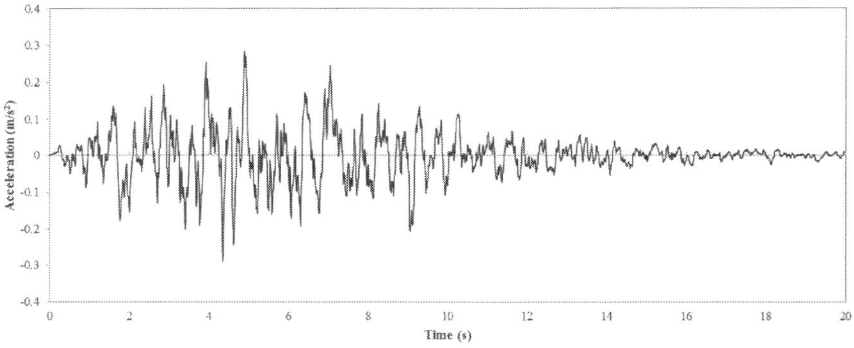

Figure 4. Artificial accelerogram.

For the analysis, artificial spectrum-compatible accelerograms have been used (**Figure 4**). It is noted that the use of real accelerograms and spectrum-matching techniques, together with records selection tools, tends to be recommended for the derivation of suits of records for use in non-linear dynamic analysis of structures but in this case, the access to real accelerograms was challenging. In literature [3], it is shown that the structural response estimated by using simulated records generally matches the response obtained using recorded motions. The software can generate artificial time-histories of ground acceleration compatible with the target spectrum and gives a comparison between its response spectrum and the target spectrum itself (**Figure 5**).

The computational analysis has been carried out using finite-element method. The model of the viaduct is three-dimensional (3D) type (**Figure 6**), representing the stiffness features of the structural elements. The composing elements are all of linear elastic type, except for the restraint devices and for the base of the piers, where it is assumed that plastic hinges may form. It takes into account geometric and material non-linearity.

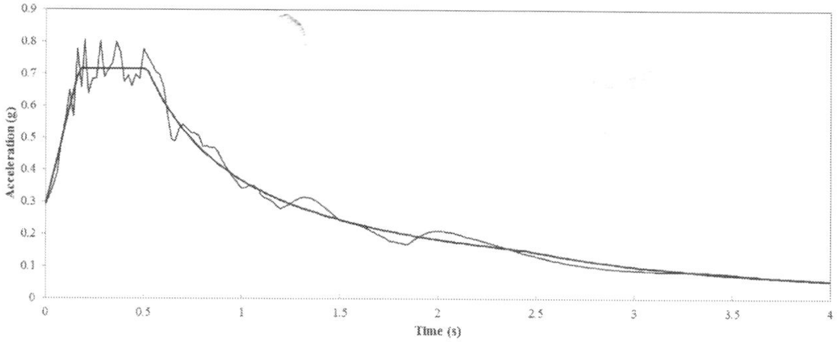

Figure 5. Compatibility between response spectrum and target spectrum.

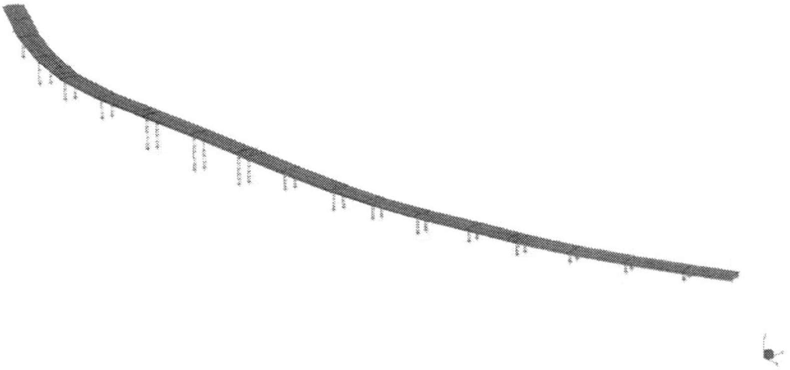

Figure 6. Finite-element model.

For the piers, a moment-curvature constitutive law has been adopted by using the Takeda model [4]. The model represents the hysteretic features of reinforced concrete structures by means of the trilinear relation force-displacement, where the non-linearity is modelled by using concentrated plastic hinges and takes into account cracking and yielding. From the force-displacement law, it is possible to retrieve the corresponding moment-curvature link. Such assessment has been carried out distinctly for each pier. With this choice, it is possible to assess in the transient state the stiffness change of the substructures, checking the consequent re-distribution of the stresses among the following piers of the viaduct.

The support and restraint devices are represented by special elements where it is possible to specify stiffness in any of six degrees of freedom between two nodes. They do not behave like a standard beam element; the degrees of freedom are user specified and are independent of each other. For non-linear analysis, each stiffness value may be defined via a non-linear table: force-velocity type for hydraulic devices and force-displacement type for isolators.

The foundations of the piers and of the abutments are modelled with punctual restraints.

2.1. Restraint systems analysis

It follows a description of the three solutions of seismic protection that will be analysed and compared in the next paragraphs.

For every solution, in order to achieve a monolithic behaviour of the deck during the seismic event, the installation of shock transmitters at the Gerber saddle provides to hinder the relative movements of the box girders. Being a dynamic device, the shock transmitters allow slow displacements (thermal change) and represent a stiff restraint against dynamic actions. The new system remains unchanged in the static stage and becomes unique kinematic chain during the seismic stage.

The restraint systems analysed original scheme, fluid viscous dampers (explained in Section 2.1.1), curved surface sliding isolators (commented in Section 2.1.2) and elastomeric isolators (described in Section 2.1.3) and are outlined in **Figure 5**.

When the viscous damper is indicated, the number following the device type represents the force threshold at which the hydraulic system starts laminating the fluid (**Figures 7** and **8**).

Figure 7. Bearing and restraint system.

		STATIC ALLOWED DISPLACEMENTS		SEISMIC ALLOWED DISPLACEMENTS	
		dL	dT	dL	dT
•	FIXED TYPE BEARING	NO	NO	NO	NO
	MULTIDIRECTIONAL TYPE BEARING	YES	YES	YES	YES
	UNIDIRECTIONAL TYPE BEARING	NO	YES	NO	YES
	UNIDIRECTIONAL TYPE BEARING	YES	NO	YES	NO
OP	VISCOUS DAMPER	NO		CONTROL	
OTP	VISCOUS DAMPER	YES		CONTROL	
OT(P)	VISCOUS DAMPER	THERMAL ONLY		CONTROL	
OT	SHOCK TRANSMITTER	THERMAL ONLY		NO	
	CURVED SURFACE SLIDER	CONTROL	CONTROL	CONTROL	CONTROL
n	ELASTOMERIC ISOLATOR	CONTROL	CONTROL	CONTROL	CONTROL

Figure 8. Map legend.

2.1.1. Hydraulic dissipation devices

The system consists of a multidirectional-encapsulated neoprene support coupled with different types of hydraulic viscous devices. They work at a speed range compatible with those of the target seismic event.

When the system reaches the design velocity that corresponds to a force threshold (following the exponential law at Eq. (2)), the piston starts to laminate the silicone fluid and it allows increasing displacements while the force level, transferred to the substructure, is kept constant.

The oleodynamic-plastic (OP) devices provide a stiff restraint against static forces but dissipate the seismic energy. It means that they can control the force level transferred to the piers up to a design limit (about 100/150 tons).

The oleodynamic-thermal-plastic devices (OTP) permit thermal expansions without remarkable resistance; instead, during the seismic stage, they are able to dissipate energy above an imposed strength level (work as an OP for dynamic forces).

The OT(P) devices are able to act as an OP (providing stiffness and dissipating energy) for high-speed displacements, induced by impulsive forces of dynamic (earthquake) and static (wind and braking) source. They behave like an OTP for low-speed displacements (temperature changes) that are allowed. Its benefit is distributing operation forces, such as braking, on a greater number of piers, reducing the forces of the fixed central device of the train.

The constitutive equation force-velocity, which characterises them, is non-linear:

$$F = C \cdot v^{a} \tag{2}$$

where C is the damping constant, v is the velocity and α is variable from 0.10 to 0.15, according to the device.

The OT (hydraulic-thermal) devices, finally, provide for a stiff restraint during the seismic event, acting as a fixed device, while allowing low-speed displacements in the static stage (thermal expansion).

All the parameters, used in the analysis, are specified in **Chart 1** for each device.

	Ab.A	P1	P2	P3	P4	P5	P6	P7	P8	P9	
type	1	3	5	4	4	4	4	4	5	5	
kH	7.04	9.82	11.88	14.14	14.14	14.14	14.14	14.14	11.88	11.88	*kN/mm*
kV	4877	6381	8289	9602	9602	9602	9602	9602	8289	8289	*kN/mm*
V (seismic)	10260	16910	21990	22700	22700	22700	22700	22700	21990	21990	*kN*
F (ULS)	14990	22590	27460	28700	28700	28700	28700	28700	27460	27460	*kN*

	P10	P11	P12	P13	P14	P15	P16	P17	Ab.B	
type	3	2	2	2	2	2	2	2	1	
kH	9.82	8.25	8.25	8.25	8.25	8.25	8.25	4.71	7.04	*kN/mm*
kV	6381	5517	5517	5517	5517	5517	5517	5517	4877	*kN/mm*
V (seismic)	16910	15290	15290	15290	15290	15290	15290	15290	10260	*kN*
F (ULS)	22590	21220	21220	21220	21220	21220	21220	21220	14990	*kN*

Chart 1. Features of the elastomeric isolators used in the analysis. k_H is the horizontal stiffness, k_V is the vertical stiffness, V is the maximum vertical load at load combinations including the seismic action and F is the maximum vertical load at non-seismic load combination at ULS.

2.1.2. Curved surface sliding isolators

The curved surface sliders are sliding isolators based on the working principle of the simple pendulum. In a structure isolated by means of this device, the period of oscillation mainly depends on the radius of curvature of the curved sliding surface and it is almost independent from the mass of the structure.

The energy dissipation is provided by the friction, which develops during the sliding, and the re-centring capacity is provided by the curvature of the sliding surfaces.

These devices are characterised by two concave surface sliders whose radius of curvature impose the period of oscillation and accommodate for horizontal displacements and rotations. A special thermoplastic material coupled with stainless steel is used on both primary and secondary sliding surfaces to govern the friction.

The material, an ultra-high molecular weight polyethylene, grants an optimal behaviour in terms of load-bearing capacity, friction coefficient and consequently energy dissipation, durability and stability to hysteretic displacement cycles. It is used without lubrication on the primary sliding surface, while it is dimpled and lubricated on the secondary one.

The above-mentioned devices feature a maximum vertical load equal to N_{Ed}=2200*t* and allow displacements of ± 150mm, in the load combination that includes the seismic action or foresees a horizontal displacement.

	Ab.A	P1	P2	P3	P4	P5	P6	P7	P8	P9
Long.	OTP 150	OTP 150	Fix	Fix	Fix	Fix	OT	Fix	Fix	OT
F	1500	1000					1700			1700
v	550	550								
α	0.15	0.15								
Transv.	Fix	OP150	OP100	OP150	OP150	OP150	Fix	Fix	Fix	OP100
F		1500	1000	1500	1500	1500				1000
v		550	550	550	550	550				550
α		0.10	0.10	0.10	0.10	0.10				0.10

	P10	P11	P12	P13	P14	P15	P16	P17	Ab.B
Long.	OT	OTP100	OP150	OT(P)150	OT(P)100	OT(P)150	OP150	OT(P)150	OTP150
F	1700	1000	1500	1500	1000	1500	1500	1500	1500
v		550	550	550	550	550	550	550	550
α		0.15	0.10	0.10	0.10	0.10	0.10	0.10	0.15
Transv.	OP100	OP100	OP100	OP100	OP100	OP150	OP150	OP150	Fix
F	1000	1000	1000	1000	1000	1500	1500	1500	
v	550	550	550	550	550	550	550	550	
α	0.10	0.10	0.10	0.10	0.10	0.10	0.10	0.10	

Chart 2. Features of the viscous dampers used in the analysis.

The value of the minimum dynamic friction coefficient matching with the maximum vertical design load N_{Ed} of the isolator is equal to 2.5% and varies with the vertical load, acting on the isolator.

The constitutive equation force-displacement, which characterises them, is bilinear type:

$$F = F_0 + k_r \cdot d = \mu \cdot N_{Sd} + (N_{Sd} / R) \cdot d \tag{3}$$

where μ is the friction coefficient, $N_{Sd}$is the load acting on the isolator (quasipermanent load), R is the equivalent radius of curvature, and d is the displacement.

All the parameters, characterising the devices, are specified in **Chart 2** since the vertical load on each support is different, being a continuous deck.

2.1.3. Elastomeric isolators (non-dissipating)

This kind of isolators consists in a rubber bearing made up of layers of elastomer.

These devices are characterised by low horizontal stiffness, high vertical stiffness and negligible damping capacity. These characteristics permit to increase the fundamental period of vibration of the structure and to resist to vertical loads without appreciable settling.

The fundamental design parameters used to determine vertical and horizontal stiffness are the geometrical characteristics (overall dimensions and thickness) and the mechanical characteristics of the elastomeric compound that is characterised by an effective dynamic shear modulus G equal to 1.4 MPa.

	Ab.A	P1	P2	P3	P4	P5	P6	P7	P8	P9	
N_{Ed}	22000	22000	22000	22000	22000	22000	22000	22000	22000	22000	*kN*
N_{sd}	8033	18294	21428	23447	23024	23095	23246	22452	21374	22198	*kN*
Friction type	2.5	2.5	2.5	2.5	2.5	2.5	2.5	2.5	2.5	2.5	
μ	5.792	2.916	2.556	2.500	2.500	2.500	2.500	2.500	2.561	2.500	
R	1800	1800	1800	1800	1800	1800	1800	1800	1800	1800	*mm*
d	150	150	150	150	150	150	150	150	150	150	*mm*
Kr	4463	10163	11904	13026	12791	12830	12914	12473	11875	12332	*kN/mm*
Ke	7.6	13.7	15.6	16.9	16.6	16.7	16.8	16.2	15.5	16.0	*kN/mm*
ξe	26.1%	16.5%	14.9%	14.7%	14.7%	14.7%	14.7%	14.7%	15.0%	14.7%	
Fo	465	533	548	586	576	577	581	561	547	555	*kN*
Fmax	1135	2058	2333	2540	2494	2502	2518	2432	2329	2405	*kN*

	P10	P11	P12	P13	P14	P15	P16	P17	Ab.B	
N_{Ed}	22000	22000	22000	22000	22000	22000	22000	22000	22000	*kN*
N_{sd}	19009	14576	16371	16079	15462	15612	15577	15583	5635	*kN*
Friction type	2.5	2.5	2.5	2.5	2.5	2.5	2.5	2.5	2.5	
μ	2.824	3.524	3.199	3 247	3.355	3.328	3.334	3.333	7.786	
R	1800	1800	1800	1800	1800	1800	1800	1800	1800	*mm*
d	150	150	150	150	150	150	150	150	150	*mm*
Kr	10561	8098	9095	8933	8590	8673	8654	8657	3130	*kN/mm*
Ke	14.1	11.5	12.6	12.4	12.0	12.1	12.1	12.1	6.1	*kN/mm*
ξe	16.1%	18.9%	17.7%	17.9%	18.3%	18.2%	18.2%	18.2%	30.8%	
Fo	537	514	524	522	519	520	519	519	439	*kN*
Fmax	2121	1728	1888	1862	1807	1821	1817	1818	908	*kN*

Chart 3. Features of the sliding isolators used in the analysis. K_r is the restoring stiffness, K_e is the equivalent stiffness (in case of linear analysis), ξ_e is the effective viscous damping, F_0 is the friction force developed by the isolator and F_{max} is the maximum horizontal force.

The compounds contain suitable antiageing additives that guarantee limited variation of the physical and mechanical characteristics in time.

Their behaviour is modelled as linear by means of the equivalent stiffness. Five types of isolators have been defined. All the parameters, characterising the devices, are specified in **Chart 3** since the vertical load on each support is different, being a continuous deck.

2.2. Non-linearity considered

According to the standards [1, 2], the structure needs to be designed in order to develop a stable dissipative mechanism, where dissipation is attributed to the piers (plastic hinges) or to appropriate devices. The elements not involved in the dissipative process need to remain elastic similar to the substructure of isolated bridges.

In this retrofit, the dissipation is committed to devices, and so it is necessary to check the behaviour of the base section of the piers.

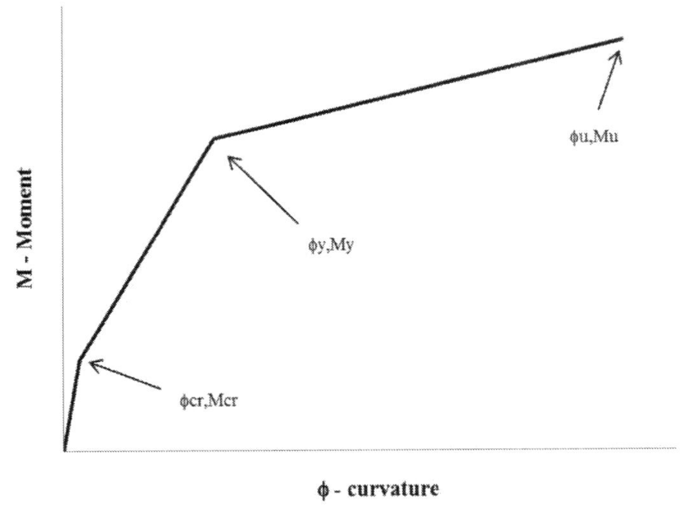

Figure 9. Constitutive diagram of the piers.

The hysteretic model of Takeda has been selected to represent the non-linear behaviour of the concrete structures. It is a realistic conceptual model that recognises the continually varying stiffness and energy-absorbing characteristics of the structure. Three linear segments define the primary curves (**Figure 9**), and the first break refers to cracking. The yield load (end point of the second line) is obtained assuming a parabolic compressive stress-strain curve for the concrete. The yield deflection depends on the deflection caused by curvature based on cracked section, the shearing deflection and the deflection caused by slip of the reinforcement and depression of the concrete. The slope of the third segment is related to the strain-hardening properties of the reinforcement. The response under load reversal, depending on the displacement ductility and the axial load ratio, is explained in the study of Takeda et al. [4].

The dissipating devices have been modelled as springs and they have been characterised by their constitutive law, described in the previous paragraphs, that is force-velocity or force-displacement type (as shown in **Figure 10**).

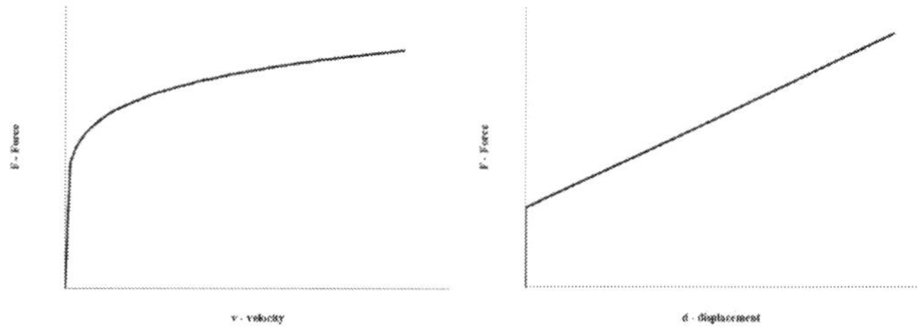

Figure 10. Constitutive laws of dissipative devices, dampers and sliders.

The following charts recall the features of each device of the restraint diagrams outlined in **Figure 5** and described in Sections 2.1.1 and 2.1.2 (**Chart 4**).

	Pier 1		Pier 2		Pier 3		Pier 4		Pier 5	
	φ[1/m]	M [kNm]	φ[1/m]	M [kNm]	φ[1/m]	M [kNm]	φ[1/m]	M [kNm]	φ[1/m]	M [kNm]
failure	4.11E-03	31000	3.24E-03	43000	2.33E-03	49000	2.57E-03	44000	1.81E-03	42000
yielding	1.37E-03	26000	1.29E-03	37000	1.45E-03	45000	1.51E-03	40000	1.50E-03	38000
cracking	2.35E-04	10000	2.70E-04	16000	3.14E-04	20000	3.34E-04	18000	3.13E-04	17000

	Pier 6		Pier 7		Pier 8		Pier 9		Pier 10	
	φ[1/m]	M [kNm]	φ[1/m]	M [kNm]	φ[1/m]	M [kNm]	φ[1/m]	M [kNm]	φ[1/m]	M [kNm]
failure	2.50E-03	54000	2.50E-03	54000	2.90E-03	50000	2.85E-03	35000	3.38E-03	36000
yielding	1.33E-03	49000	1.33E-03	49000	1.26E-03	44000	1.36E-03	31000	1.31E-03	31000
cracking	2.83E-04	22000	2.83E-04	22000	2.59E-04	19000	2.56E-04	13000	2.53E-04	13000

	Pier 11		Pier 12		Pier 13		Pier 14		Pier 15	
	φ[1/m]	M [kNm]	φ[1/m]	M [kNm]	φ[1/m]	M [kNm]	φ[1/m]	M [kNm]	φ[1/m]	M [kNm]
failure	4.14E-03	27000	4.14E-04	27000	4.06E-03	25000	4.06E-03	25000	3.29E-03	22000
yielding	1.44E-03	24000	1.44E-03	24000	1.41E-03	22000	1.41E-03	22000	1.53E-03	20000
cracking	2.78E-04	10500	2.78E-04	10500	2.97E-04	10000	2.97E-04	10000	2.73E-04	8000

	Pier 16		Pier 17	
	φ[1/m]	M [kNm]	φ[1/m]	M [kNm]
failure	3.29E-03	22000	3.29E-03	22000
yielding	1.53E-03	20000	1.53E-03	20000
cracking	2.73E-04	8000	2.73E-04	8000

Chart 4. Parameters of non-linear material characterising the constitutive law of the pier bases.

3. Results

The results of the analyses performed have been presented as a comparison among the restraint diagrams proposed, in terms of stresses at the base of the piers (**Figures 11** and **12**), displacements on the top of the pier (**Figure 13**), displacements and forces acting on the devices

(**Figures 14** and **15**) and dissipated energy (**Figures 16** and **17**). The following charts relate to the maximum stresses for each base section of the piers. These results were assessed as the square root of the instant-by-instant stresses, since the pier is circular; the charts also show the values of the "yield moment" of the examined sections.

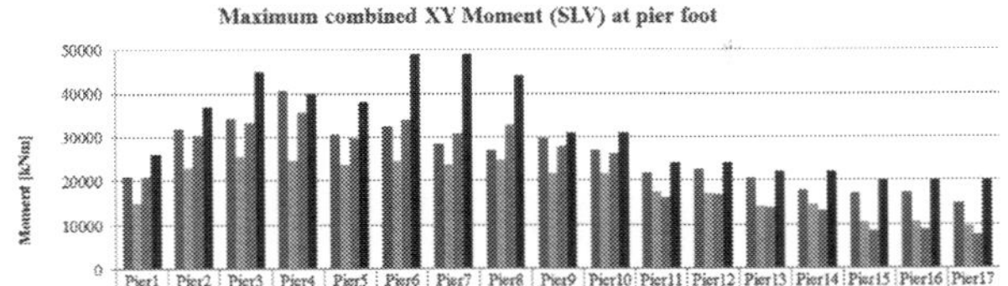

	Pier1	Pier2	Pier3	Pier4	Pier5	Pier6	Pier7	Pier8	Pier9	Pier10	Pier11	Pier12	Pier13	Pier14	Pier15	Pier16	Pier17
Viscous dampers	20914	31783	34156	40607	30567	32378	28463	26840	29623	26883	21764	22539	20475	17847	17110	17255	14969
Curved surface sliders	15059	23000	25399	24484	23656	24440	23573	24680	21612	21580	17422	17015	14174	14799	10611	10635	9584
Elastomeric isolators	20902	30455	33271	35576	29621	33943	30721	32694	27794	26217	16107	16905	13980	13260	8573	8921	7772
Yeld limit	26000	37000	45000	40000	38000	49000	49000	44000	31000	31000	24000	24000	22000	22000	20000	20000	20000

Figure 11. Bar chart of the bending stresses ULS (max XY) at the base of the piers.

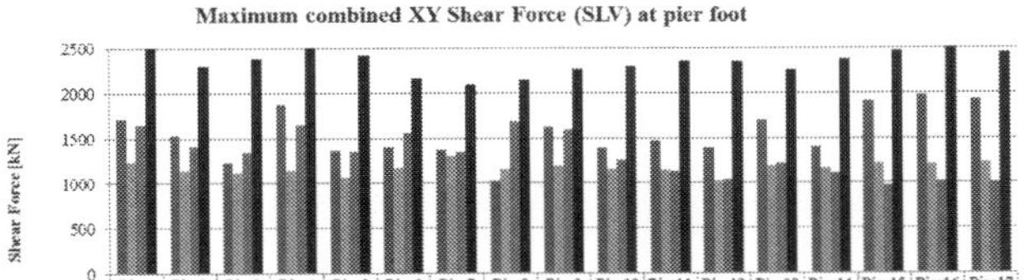

	Pier1	Pier2	Pier3	Pier4	Pier5	Pier6	Pier7	Pier8	Pier9	Pier10	Pier11	Pier12	Pier13	Pier14	Pier15	Pier16	Pier17
Viscous dampers	1711	1524	1227	1874	1368	1400	1372	1032	1619	1392	1463	1389	1698	1404	1907	1971	1924
Curved surface sliders	1235	1142	1115	1149	1061	1177	1304	1151	1188	1153	1141	1024	1187	1168	1214	1211	1224
Elastomeric isolators	1642	1406	1348	1648	1354	1557	1347	1679	1589	1253	1131	1041	1217	1108	977	1014	1008
Shear Resistent	2506	2300	2379	2509	2416	2156	2089	2144	2264	2288	2349	2342	2252	2366	2457	2531	2440

Figure 12. Bar chart of the shear stresses ULS at the base of the piers.

As asserted in Section 2.2, the main requirements are to ensure that the substructure behaves elastically.

It can be deduced from **Figure 11** how all the solutions lead to stress levels that are lower than the yield limit of the piers. Furthermore, the solution with viscous devices better exploits the bending-resistant capacity and the shear-resistant capacity of the reinforced concrete section.

The following chart refers to the displacement of pier top. This parameter is linked to the control of the local ductility at the plastic hinges [5] by the chord rotation. The latter is measured over the length of the pier, between the end section of the plastic hinge and the section of zero moment.

The verification that deformation demands are safely lower than the capacities of the plastic hinges should be performed by comparing plastic hinge rotation demands, $\theta_{p,E}$, to the relevant design rotation capacities, $\theta_{p,d}$, as follows: $\theta_{p,E}<\theta_{p,d}$.

Again, it is necessary to ensure that the displacements are lower than the yield displacement limit (**Chart 5**):

$$D_y = \phi_y \cdot H^2 / 3 \tag{4}$$

The following charts refer to devices, and in particular, they show displacements and the conveyed forces.

The chart in **Figure 14** regards the displacements of the devices and to be understood it needs to be read together with the chart in **Figure 15**, which concerns about the stress level reached by the devices.

In general, the amount of the displacements increases when the height of the pier is low, due to the ductility of the element. Considering **Figure 13**, the displacement of the pier is greater for higher piers.

It is possible to affirm that for lower piers (from pier 11 to 17), the behaviour of sliders and elastomeric isolators in terms of strength is similar; while the displacement is different, the latter request more displacement than the former in order to guarantee the same level of force.

The displacements of OP-type dampers are null, and from the bar chart shown in **Figure 13**, it is possible to appreciate that the displacement of the top of the piers with fix restraint or OP restraint is greater.

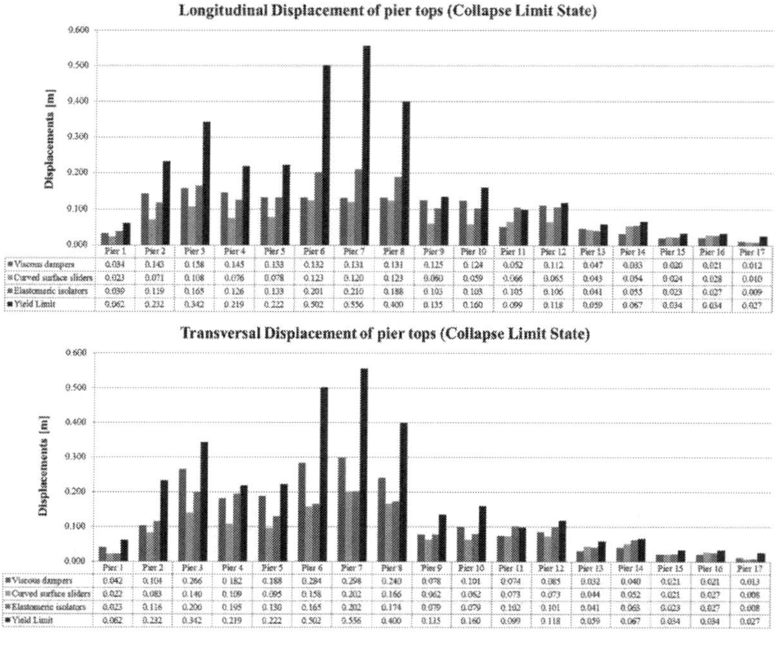

Figure 13. Bar charts representing the displacements on the level of the heads of the piers.

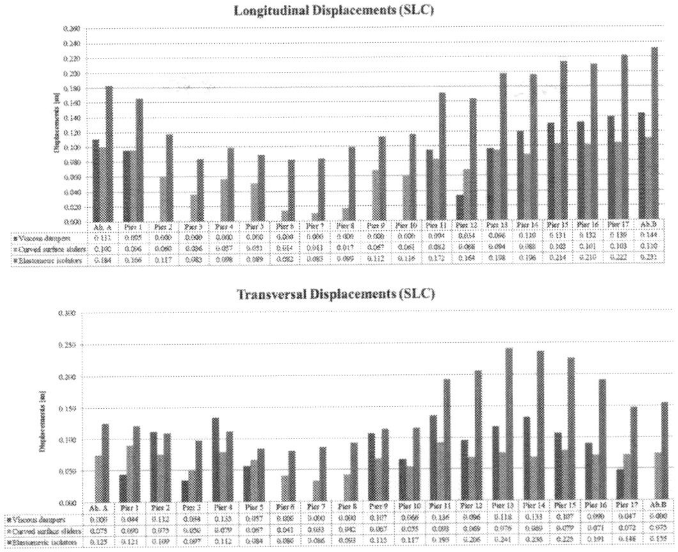

Figure 14. Bar charts representing the displacements on the level of the devices.

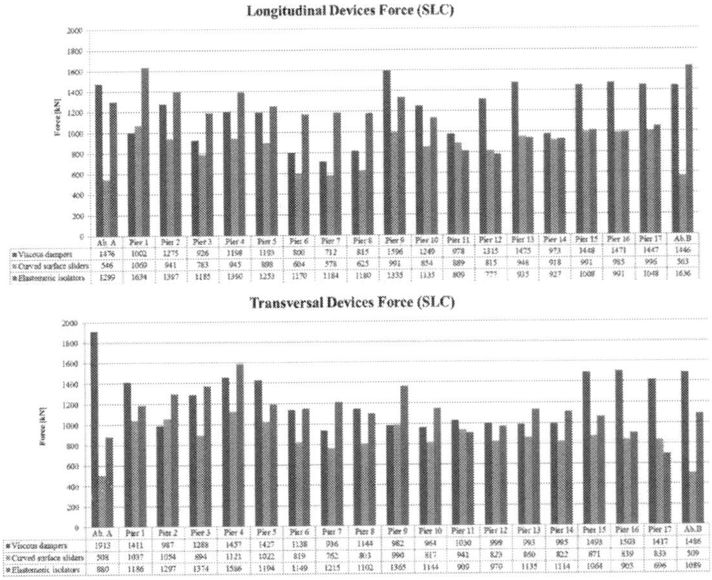

Figure 15. Bar charts representing the forces on the devices.

The following diagrams contain the hysteretic energy dissipation processes of the following two main opposing mechanisms of the earthquake (devices and piers): the devices in terms of force-displacement (there is dissipation only in case of dampers and sliders) and the material of the piers, relating to the base section, in terms of moment-curvature. The area underlying the curves is an indicator of the dissipated energy.

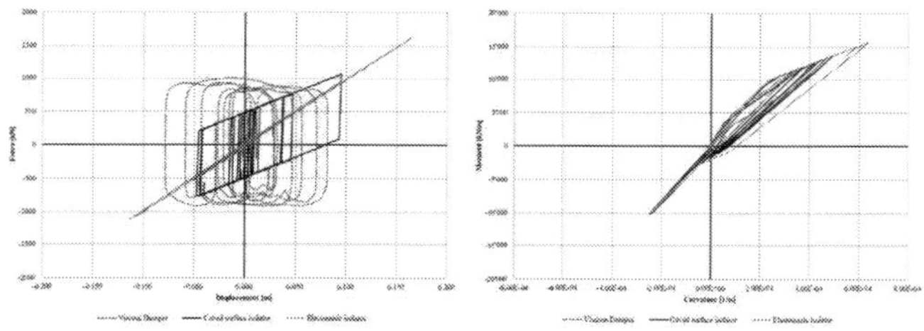

Figure 16. Comparison of energy dissipation (restraint device and section at the base of the pier n.1 – earthquake ULS of Collapse).

The following chart (**Figure 17**) summarises the endeavour performed by the piers during the duration of the earthquake: the same is obtained as the sum of the products of forces at the base of the pier and the displacement at the top of the same pier for each moment of the seismic event. In particular, it is clear how the piers are implicated in the total dissipation process of the seismic energy, partly carried out also by the devices in terms of displacement. Moreover, it can be deduced how the solution with the sliders allows transferring only a reduced portion of energy to the piers.

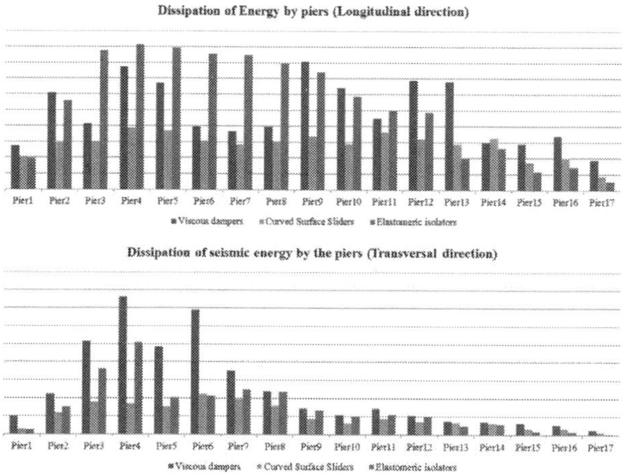

Figure 17. Bar charts representing energy dissipation in the earthquake expressed by the piers.

http://dx.doi.org/10.5772/63821

	ϕy	H	θy	Dy
Pier 1	1.37E-03	12.75	5.82E-03	0.062
Pier 2	1.29E-03	25.45	1.09E-02	0.232
Pier 3	1.45E-03	29.15	1.41E-02	0.342
Pier 4	1.51E-03	22.85	1.15E-02	0.219
Pier 5	1.50E-03	23.10	1.16E-02	0.222
Pier 6	1.33E-03	36.85	1.63E-02	0.502
Pier 7	1.33E-03	38.80	1.72E-02	0.556
Pier 8	1.26E-03	33.80	1.42E-02	0.400
Pier 9	1.36E-03	18.90	8.57E-03	0.135
Pier 10	1.31E-03	21.00	9.17E-03	0.160
Pier 11	1.44E-03	15.70	7.54E-03	0.099
Pier 12	1.44E-03	17.20	8.26E-03	0.118
Pier 13	1.41E-03	12.30	5.78E-03	0.059
Pier 14	1.41E-03	13.05	6.13E-03	0.067
Pier 15	1.53E-03	8.90	4.54E-03	0.034
Pier 16	1.53E-03	8.90	4.54E-03	0.034
Pier 17	1.53E-03	7.90	4.03E-03	0.027
	[1/m]	[m]	[rad]	[m]

Chart 5. Displacements correspondent to yield limit curvature of the pier.

4. Conclusion

The analysis focused on three different retrofit solutions of an existing viaduct, examining the technical aspects relating to the design of dissipation and isolation systems. When it is necessary to operate on existing structures, a special attention and carefully conceived solutions are mandatory. Regarding the actual design of the viaduct in question, the restraints are related to the impossibility of stopping traffic and to intervening by reinforcing the existing piers, respecting limitations imposed by the client regarding safety conditions during the implementation of the works.

4.1. General considerations

All the solutions proposed have been designed and calibrated in order to respect the requirement of not using the plastic reserve of the piers, to contrast the forces induced by the system. The refining has related to the modulation of the operative velocity of the viscous devices, the definition of the curvature of the sliding isolators and the level of the stiffness of the rubber of the elastomeric isolators (as reported in **Charts 1–3**). The chart, in **Figure 10**, represents the comparison between the stressing actions and the yield limit of the substructures.

4.2. Solution with fluid dynamic dissipation devices (Sol. 1)

The solution respects all the restraints desired by the client. It is not necessary to intervene by reinforcing the piers, the gaps on the abutments are assured (displacements less than 15 cm), the bearing system of the current situation even in the transient phase (during works) is assured and the structure is totally verified. Implementing this solution requires rather high implementation costs. In fact, it is necessary to remodel the top of each pier by widening the top to allow the housing of the contrast structures of the dissipating devices. Another relevant cost can be ascribed to the assembly of the same dissipating devices.

From a global point of view (not strictly engineering), this is the best solution as it allows a check of the stresses acting on the substructures, regardless of the intensity of the seismic event, at the expense of the displacement of the devices. The devices are provided by law with a safety margin to be able to tackle greater displacements compared to those for which they have been designed. Therefore, it is necessary to implement the joint gaps with a consistent safety margin in order to avoid hammering between the deck and the abutment structures.

4.3. Solution with curve sliding isolators (Sol. 2)

The solution with the sliders is the most efficient, in regards to the stresses on the substructures and for the displacements during the seismic stage. But it does feature some issues in regards to the management of the bearing system during assembly, given that the actual fixed points of the deck are removed. During operation, although, for horizontal loads, due to vehicular traffic (braking) and to wind, the minimum friction allows to develop a breakaway torque (F_0) greater than the one, to which each support is subject, thermal expansion entails displacement at all the piers. This solution also allows a low implementation cost, since the support integrated the isolator, which features nearly the same sizes of the existing support.

From an engineering point of view, the solution is efficient and allows to involve in the dissipation process, both the piers (within the allowed capacity) and the devices (owing to the friction developed by the material interposed between the curved surfaces), with a work rate lower than the one required by the solution with only the dampers. Since the sliders is at the same time an isolator, the solution, moreover, allows enhancing the total response of the structure, defining a new oscillation period, independent from the deck mass but depending on the curvature of the devices. With reference to the modal analysis with the project spectrum, the structure undergoes a smaller seismic acceleration at the base, due to an increase in the period.

As far as realisation is concerned, the solution shortens the supply period (same type of device on all the piers), simplifies the launching operations for implementation, as the new devices feature sizes and space requirements similar to the existing supports. Yet, the type of support is not suitable for this specific viaduct, featuring a central point-shaped bearing. It entails the application of auxiliary temporary restraints, for the management of the transitory stage, which is not easy to implement.

4.4. Solution with elastomeric isolators (Sol. 3)

The solution with elastomeric isolators is the less suitable option in this case. Although it allows not intervening on the piers, in regards to the base and heading sections, it involves more displacements than the previous solutions (thus requiring the widening of the gaps on the abutments, thus necessarily interfering with traffic circulation). As Sol. 2, it does not preserve the bearing system of the current situation in the transitory stage as desired by the client.

Yet, this solution has been analysed and compared, since it is one of the most widespread as for this type of intervention, and it is undoubtedly the cheapest.

	Sol. 1	Sol. 2	Sol. 3
Preservation of the bearing system of the current situation in the transitory stage	✓		
Displacements viaducts abutments	✓	✓	
Stresses within elastic limit on the substructures	✓	✓	✓
No intervention on pier top		✓	✓
Easy implementation			✓
Limited tests on devices			✓
Consistency of devices		✓	
Control of forces for each seismic return period (T_R)	✓		

Author details

Alessandro Contin* and Andrea Mardegan

*Address all correspondence to: alessandro.contin@e2b.it

E2B s.r.l., Padova, Italy

References

[1] Nuove Norme Tecniche per le Costruzioni – DM Infrastrutture 14/01/2008, G.U. 4/2/2008 n.29 NTC (2008).

[2] Consiglio superiore dei lavori pubblici. Istruzioni per l'applicazione delle "Norme tecniche per le costruzioni" di cui al D.M. 14/01/2008.

[3] Galasso C., Zhong P., Zareian F., Iervolino I., Graves R.W. (2013). "Validation of ground-motion simulations for historical events using MDoF systems." Earthquake Engineering and Structural Dynamics.42(9):1395–1412.

[4] Takeda T., Sozen M.A., Norby Nielsen N.(1970). Reinforced concrete response to simulated earthquakes, Journal of the Structural Division, Proceeding of the American Society of Civil Engineers.

[5] UNI EN 1998-2:2012. Eurocode 8: Design of structures for earthquake resistance. Part 2: bridges.

Chapter 6

Fluid Viscous Dampers and Shock Transmitters in Realization of Multi-Span Steel-Concrete Viaducts

Alessandro Contin and Andrea Mardegan

Additional information is available at the end of the chapter

http://dx.doi.org/10.5772/64038

Abstract

This chapter discusses the methodology for the dissipation of seismic energy, designed for the construction of a steel-concrete viaduct in a variable-orography land. The viaduct has a total length of 1102 m and typical spans of 75 m, with piers, of a maximum height of 65 m, significantly varies from each other. The viaduct is subjected to a redesign step in order to adapt it to the requirements of Italian standard 'D.M. 14/01/2008'. The new design has reformulated the sequence of spans and, consequently, redefined the structures constituting the foundations, piles, steel girders, inferior bracings and especially the typology of bearings and seismic devices.

The hydraulic devices are therefore an efficient solution for the control of seismic stresses induced by continuous viaducts on substructures, especially when the piers of the viaduct deck themselves and ensure structural safety.

Keywords: non-linear analysis, seismic energy dissipation, fluid viscous dampers, shock transmitters, anti-seismic devices

1. Introduction

Steel-concrete composite bridges had a strong growth in railways and highways since the construction system provides an efficient and cost-effective form. The combination of steel and concrete in a single composite structural element enhances the individual advantages of both materials. By utilizing the tensile strength of steel in the main girder and the compressive strength of concrete in the slab, the bending resistance of the combined materials is greatly increased and larger spans are made possible. It allows overcoming difficulties caused by irregular orography and offers advantages in terms of resistance and ductility useful in high seismicity areas.

The chapter, precisely, concerns with a steel-concrete composite viaduct on a territory with a variable orography and focuses on the dissipation method of the seismic energy.

The context, which led to the choice of hydraulic devices with the stabilizer system, is analysed. Such devices, designed according to the stiffness of the substructures, allowed to limit and, partly, dissipate the stresses, induced by the earthquake, maintaining the underlying structures substantially elastic to the condition of LLS. Then, the checks have proven the capacity of the structures to support the stresses to limit state condition (LSC) without undergoing any damages, assuring curving capacity of the piers. The implementation of the devices has been preceded by the timely check of the functionality of the same and of the correspondence of the project-operating curve in accordance with the provisions of the pertaining European regulation, among which EN15129:2009 [1] as well as the Italian standard DM 14/01/2008 [2].

1.1. Main features of the viaduct

The viaduct covers a total length equal to 1102 m and it is made up of 15 spans, divided as follows: 68.0 + 75.0 m × 13 + 59.0 m. The piers have a maximum height of 65 m and present remarkable variability from one span to the other.

The slab is 12.0 m large and harbours a 10.5-m large lane. The main girders, 2.96 m high, have a distance of 7 m and they are linked by bracing each 6.50 m; a lower cross-bracing system provides resistance against torsional effects. This type of structure is lightweight, easy to transport and it permits easy launch.

The choice of the steel-concrete composite structure with continuous deck well adapts to the alignment and altimetric irregularities assuring reduced seismic masses at the same pace with a suitable stiffness.

The vertical structures are represented by reinforced concrete piers with pseudo-ellipsoid hollow section, with an average thickness ranging from 30 to 45 cm. Drilled piles with a diameter equal to 1.5 m, featuring a suitable resistance to side forces, in terms of both shear stress and induced bending stress, have been used for foundation structures. Such a choice is supported by the breaking strength test, carried out in the worksite, where the same piles have proven to assure and move a greater peak flow compared to the piles, having an equal length and a smaller diameter. The design has been performed in compliance with the capacity design, focusing on the plastic reserves of the structure in correspondence of the dissipation devices and of the piers (keeping the superstructure, the vertical support devices and the foundations in the elastic range).

1.2. Orography context

The viaduct is inserted in a context of deep narrow valleys, not serviced at all, and features a territory with constant lithological characteristics (although variable from a geotechnical point of view for the different compaction of the layers), where the surface of the grey-deep blue clays is from time to time covered with layers of loose yellow sand or by the relating remodelled layers of the main geological formations. Together with geotechnical irregularities, there are also some altimetric irregularities, ordered by the alignment with connection and curve areas. These irregularities affected the project choices as for both the type of construction and the earthquake-resistant system (**Figure 1**).

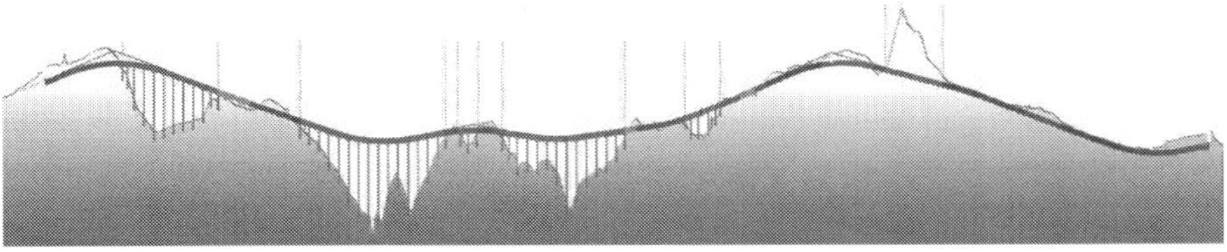

Figure 1. Orography of the site.

1.3. Choice of the earthquake-resistant system

Given the remarkable irregularity of the height of the piers and of the road alignment, it was necessary to couple the longitudinal oscillation of the deck and the oscillation of the underlying piers. Such coupling has been induced on the higher piers (having a similar stiffness) by the introduction of OT (shock-transmitter) devices. The lower piers have been left free to oscillate, whereas those with an intermediate height have also been coupled with hydraulic thermo-plastic devices with OTP viscous-type behaviour able to control the forces, conveyed to the underlying piers. Transversally, plastic hydraulic OP-type devices as well as fixed restraints have been used. Such a choice implies that the piers or the devices, alternatively, must be able to dissipate the energy of the earthquake by any formation of plastic hinges or the work expressed by the devices. In compliance with the project, with regard to seismic testing, the types of joints, able to allow the displacements during both the service and earthquakes, have been defined.

Both fluid viscous dampers and shock transmitters are integrated with bearings. They have been designed according to the different stiffness of the substructures and allow limiting and partially dissipating the stresses induced by the earthquake, keeping the deck and the substructures substantially elastic for life-safety limit state condition (at the ultimate limit state (ULS)).

The installation of the devices was preceded by the accurate analysis of their seismic behaviour and matching of functional curves with the design assumptions, in accordance with the dictates of the European Reference Standards including UNI EN15129:2009 [1] as well as the mandatory Italian Standard 'D.M. 14/01/2008' [2, 3]. Finally, dynamic tests were carried out in situ to validate the theoretical assumptions about the dynamic behaviour of the real decks and substructures (**Figures 2** and **3**).

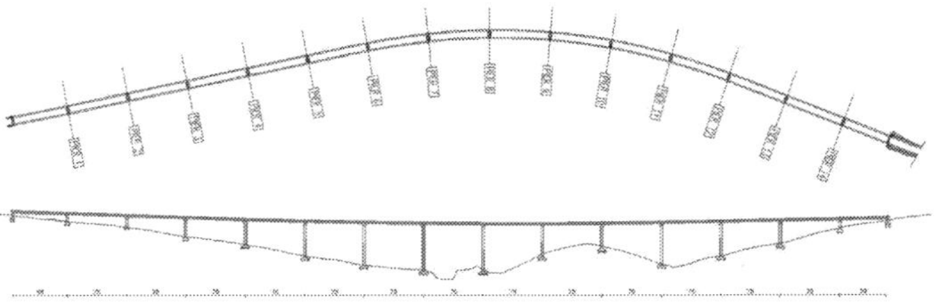

Figure 2. Viaduct plan and view.

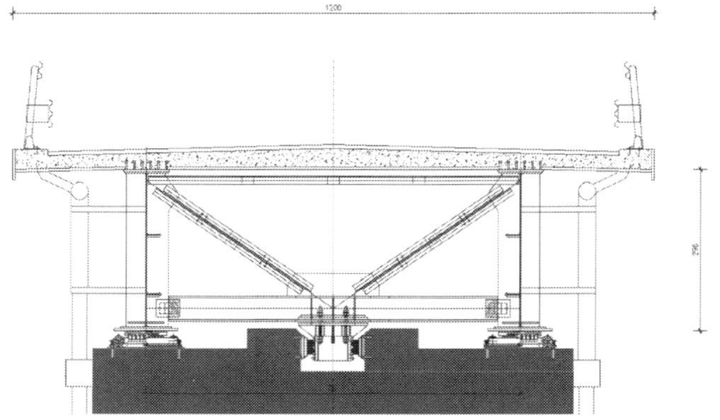

Figure 3. Viaduct cross section.

2. Modelling and design approach

The design of the viaduct was performed according to Italian code [2, 3] taking into account the dissipative behaviour of the structural elements, material and geometric non-linearity, following a high-ductility approach.

The dissipative zones are concentrated in the seismic-restraint devices while non-dissipative elements are designed according to capacity design criterion [2, 3]. The structure is conceived and designed with the goal to create a stable dissipative mechanism under the seismic action at the life-safety limit state (ultimate limit state—ULS).

The plastic deformations of the base of the piers generate displacement demand that is requested to be lower than the capacity of the system. The comparison between ductility demand and available ductility was carried out on the basis of the instructions provided in [4]. In detail, deck, vertical support devices, foundation structures and abutments are designed to remain elastic.

'Over-strength' factors were considered for the verification of the pier sections outside the plastic hinge region and for the foundations as well. The fixed constraint devices were dimensioned according to the capacity design criterion. The dissipative devices were designed to support, without collapse, earthquake displacements caused at the collapse limit state (CLS).

The computational analysis has been carried out using finite element method (FEM). The model of the viaduct is of a three-dimensional type (**Figure 4**) and recreates the stiffness of the structural elements, constituting it, the non-linear features are concentrated within the restraint devices and at the base of the piers (plastic hinges). It takes into account geometric and material non-linearity.

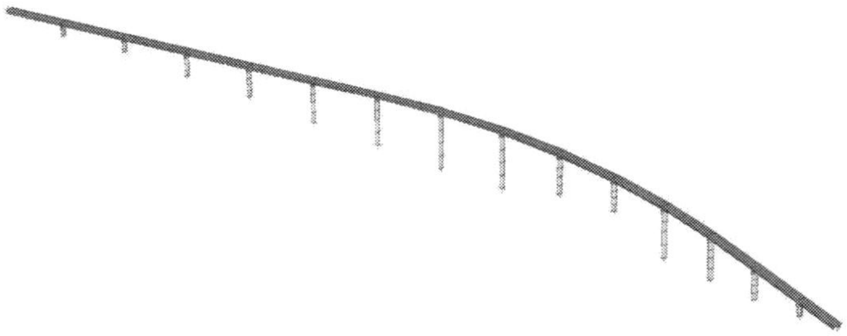

Figure 4. Finite element model.

For the piers, a moment-curvature constitutive law has been adopted by using the Takeda model [5]. The model represents the hysteretic features of reinforced concrete structures by means of the trilinear relation force displacement (**Figure 5a**), where the non-linearity is modelled by using concentrated plastic hinges and takes into account cracking and yielding. From the force-displacement law, it is possible to retrieve the corresponding moment-curvature link. Such assessment has been carried out distinctly for each pier. With this choice, it is possible to assess in the transient state the stiffness change of the substructures, checking the consequent redistribution of the stresses among the following piers of the viaduct.

The OP and OTP viscous dissipation devices are devices provided with a cylinder and a piston, depending on the velocity, where the lamination of a silicone fluid by means of a suitable hydraulic circuit allows energy dissipation. The typical constitutive law force-velocity, simulating the behaviour is of non-linear type (**Figure 5b**), is

$$F = C \cdot V^{\alpha} \tag{1}$$

where C represents the damping constant and α is assumed to be equal to 0.15.

The dissipation devices have been modelled by means of spring-damper elements with the association of the law force-velocity (Eq. (1)). There are two types of devices, a plastic OP hydraulic one and the other one is an OTP thermal-plastic hydraulic one.

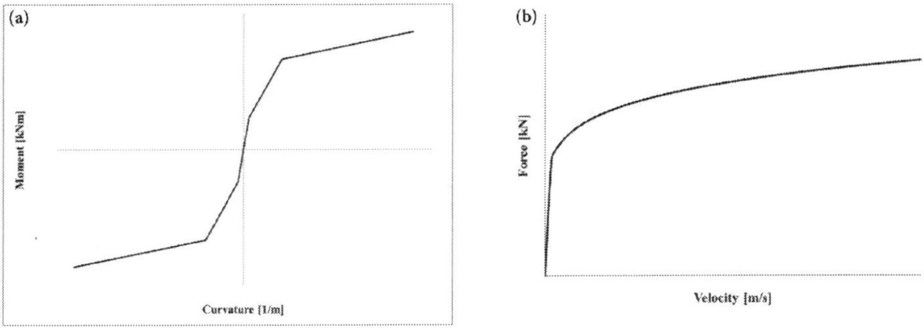

Figure 5. (a and b) Flow chart moment curvature according to Takeda and force-velocity law ($F = C \times V^{\alpha}$).

The OT devices, with a dynamic restraint (shock transmitter), represent a very stiff restraint against a dynamic action, whereas they allow slow displacements of the structures (e.g. due to thermal changes). Owing to their features, they have been modelled as truss elements with a high stiffness.

2.1. Design choices relating to restraint diagram

The restraint system is outlined in **Figure 6**. Longitudinally, the fixed piers (P7 and P8) absorb static stresses due to braking and play the role of a thermal centre point by means of a transversal one-way restraint; from a seismic point of view, the shorter piers (P–P3 and P14 with a height of less than 25 m) are free to oscillate, piers P3–P13 (about 30 m high) are provided with OTP-type dissipation devices, able to check the stress value, given by the deck; for the remaining piers (with height exceeding 35 m), the application of temporary-restraint devices (shock-transmitter OT) is provided. Transversally, the piers (P1–P4, P10, P13, P14) are provided with a multidirectional support, associated to an OP plastic hydraulic type device, whereas P5, P6, P9, P11, P12 are associated with a DEF* (*with shoes, able to accept longitudinal displacements) fixed type restraint; piers P7 and P8 are provided with a transversal one-way

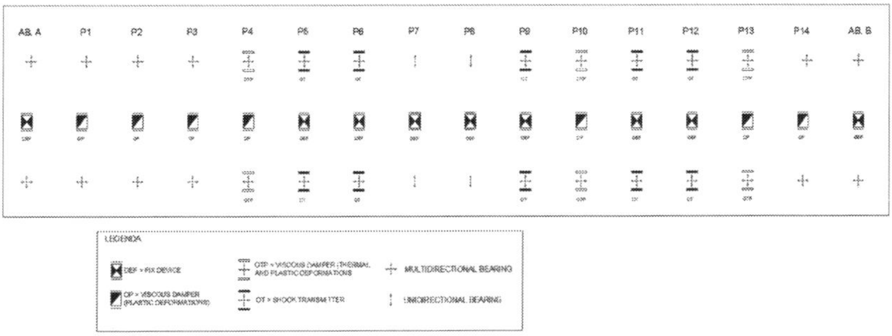

Figure 6. Restraint diagram.

http://dx.doi.org/10.5772/64038

support device, associated with a DEF-type device. From a seismic point of view, only the piers, being less than 35 m high (P1–P4, P10, P13, P14), have been isolated. The classification diagram of the piers is represented in **Figure 7**.

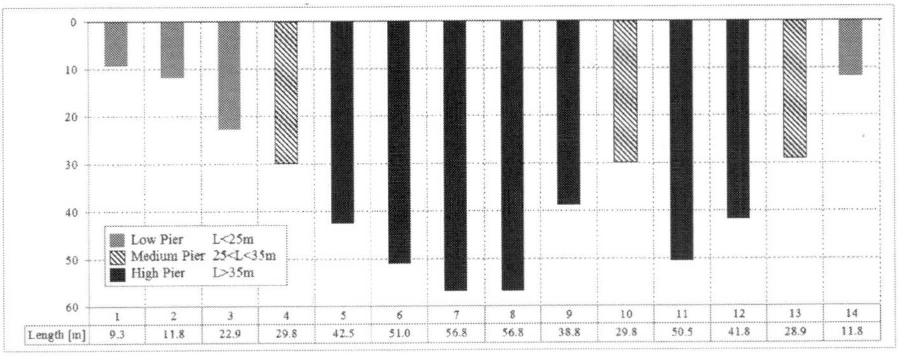

	1	2	3	4	5	6	7	8	9	10	11	12	13	14
Pier Length [m]	9.3	11.8	22.9	29.8	42.5	51.0	56.8	56.8	38.8	29.8	50.5	41.8	28.9	11.8
Length classification	LOW	LOW	LOW	MED	HIGH	HIGH	HIGH	HIGH	HIGH	MED	HIGH	HIGH	MED	LOW
Long. Device	MULTI	MULTI	MULTI	OTP	OT	OT	UNID.L	UNID.L	OT	OTP	OT	OT	OTP	MULTI
Classification	**free**	**free**	**free**	**viscous**	**fix**	**fix**	**fix**	**fix**	**fix**	**viscous**	**fix**	**fix**	**viscous**	**free**

	1	2	3	4	5	6	7	8	9	10	11	12	13	14
Pier Length [m]	9.3	11.8	22.9	29.8	42.5	51.0	56.8	56.8	38.8	29.8	50.5	41.8	28.9	11.8
Length classification	LOW	LOW	LOW	MED	HIGH	HIGH	HIGH	HIGH	HIGH	MED	HIGH	HIGH	MED	LOW
Trasv. Device	OP	OP	OP	OP	DEM	DEM	DEF	DEF	DEM	OP	DEM	DEM	OP	OP
Classification	**viscous**	**viscous**	**viscous**	**viscous**	**fix**	**fix**	**fix**	**fix**	**fix**	**viscous**	**fix**	**fix**	**viscous**	**viscous**

Figure 7. Piers classification diagram.

2.2. Analyses performed

The assessment of the actions, due to the earthquake, has been carried out by means of a non-linear dynamic analysis with step-by-step integration. The masses associated to gravitational loads have been assessed as follows:

$$G_1 + G_2 + \Sigma_j \psi_{2j} Q_j \text{ with } \psi_{2j} = 0.2 \quad (2)$$

where G_1 is the weight of the structure; G_2 is the permanent load; Q_j is the traffic load.

The procedure adopted provides for

- Non-linear step-by-step dynamic analysis (three groups of accelerograms in the three directions applied simultaneously). The effects on the structure are represented by the most adverse values of the effects obtained by the analysis by using the three different groups of accelerograms for ULS.

- Linear step-by-step dynamic analysis for comparison with non-linear step-by-step analysis in order to establish the structure factor q, implicitly deriving from the hysteretic cycle of load-unload of the materials, making up the substructures. With such q value, then, the over-strength factor, required for the check of the sections, placed outside of the critical areas, has been calculated (capacity design).
- Modal analysis with design spectrum (using the estimated q) and approval of the non-linear step-by-step analysis by comparison with the total forces at the base of the piers.

For seismic purposes, the following parameters (referred to DM 14/01/2008 [2]) have been taken into account: rated life of the works, V_N = 50 years; use coefficient of the works, C_U = 2. The result is a reference period equal to V_R = 100 years. The soil has been considered of class C: deposits of medium thickened soil, with layers of more than 30m where mechanical properties gradually increase with depth.

3. Analysis results and match with test data

The results of the seismic computational analyses are shown in the following charts. They are presented as a comparison between theoretical and experimental data. The former are characterized by the law of materials and devices deriving from mathematical assumptions, the latter instead use the output results of the laboratory-type test (**Figure 8**). The constitutive diagram of the piers, shown in **Figure 9**, remains the same in both analyses.

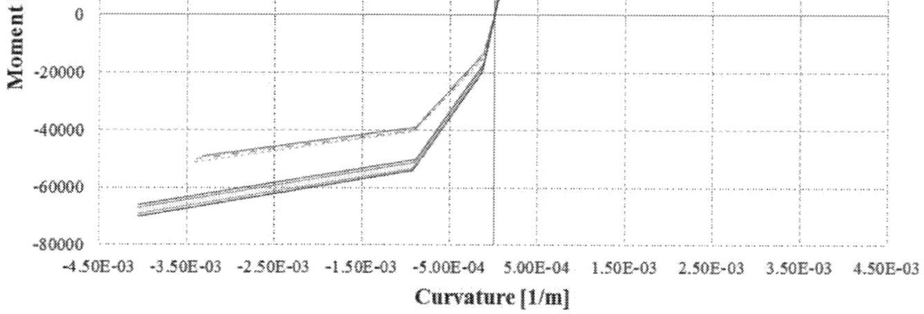

Figure 8. Comparison between theoretical and type-test law for devices.

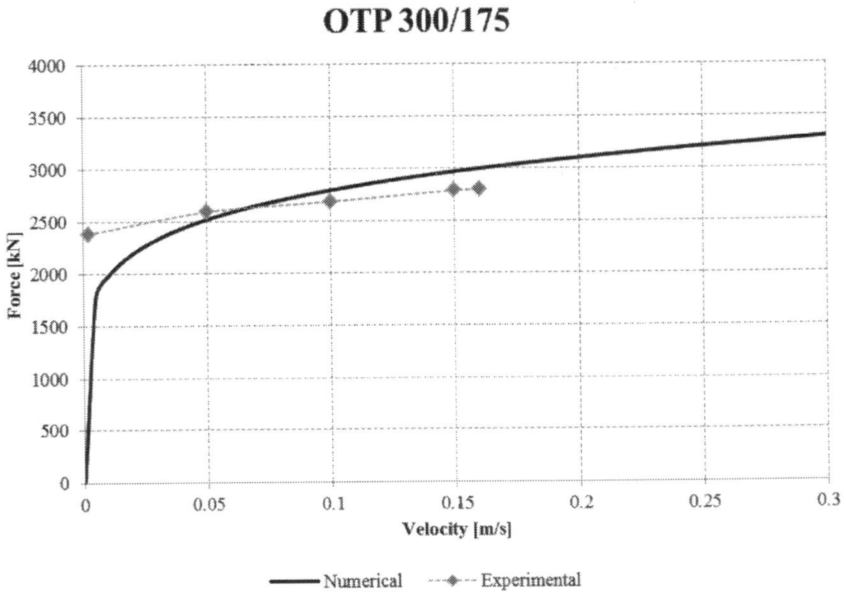

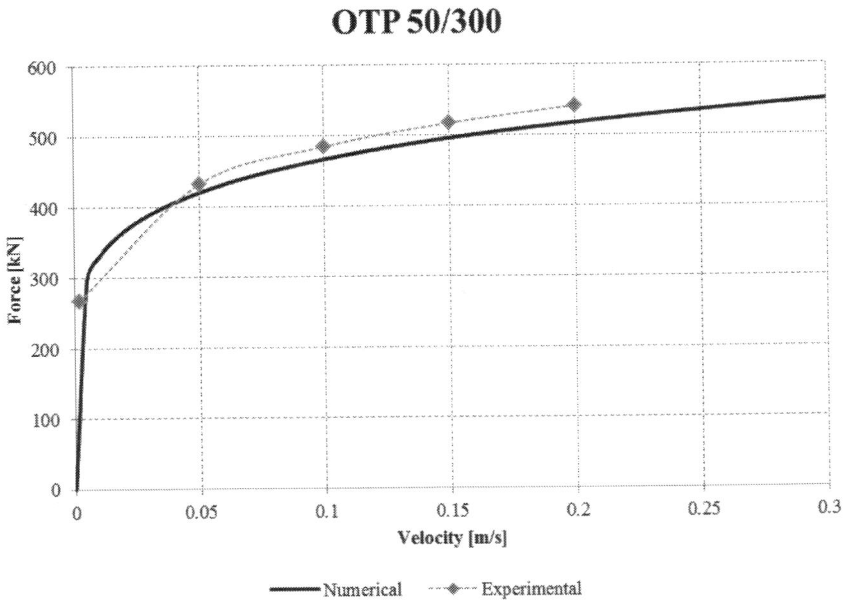

Figure 9. Constitutive diagrams of the piers.

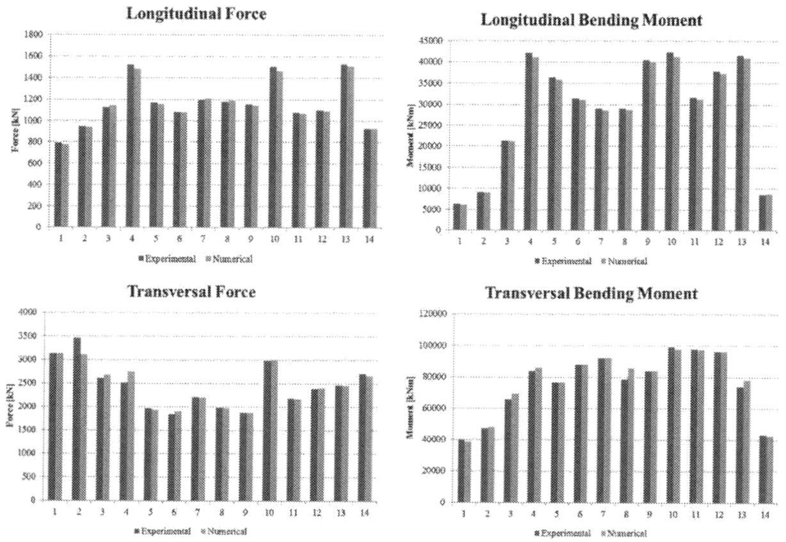

Figure 10. Representative bar charts of the forces and moments at the base of the piers—comparison between experimental and numerical results.

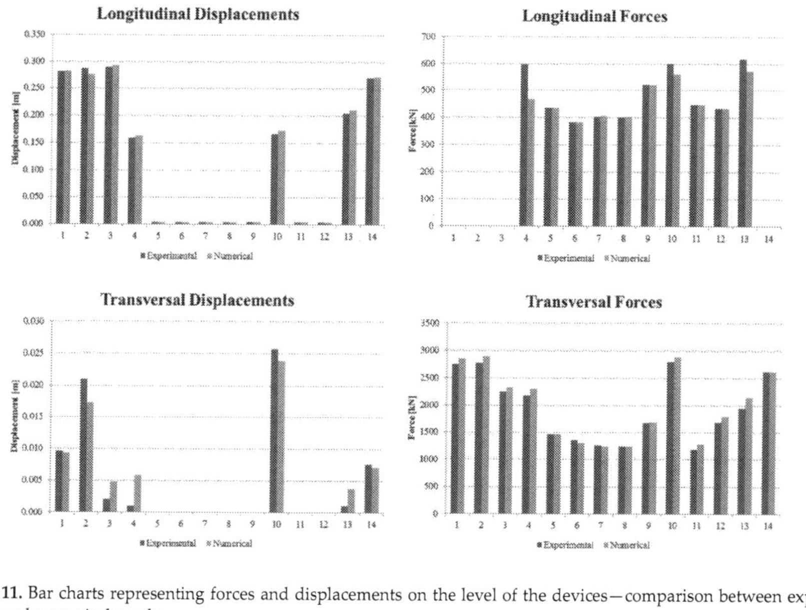

Figure 11. Bar charts representing forces and displacements on the level of the devices—comparison between experimental and numerical results.

http://dx.doi.org/10.5772/64038

The results of the non-linear analyses are represented here below; each chart shows by the most adverse effects of three accelerograms at the ULS applied as explained in Section 2.2 (**Figures 10** and **11**).

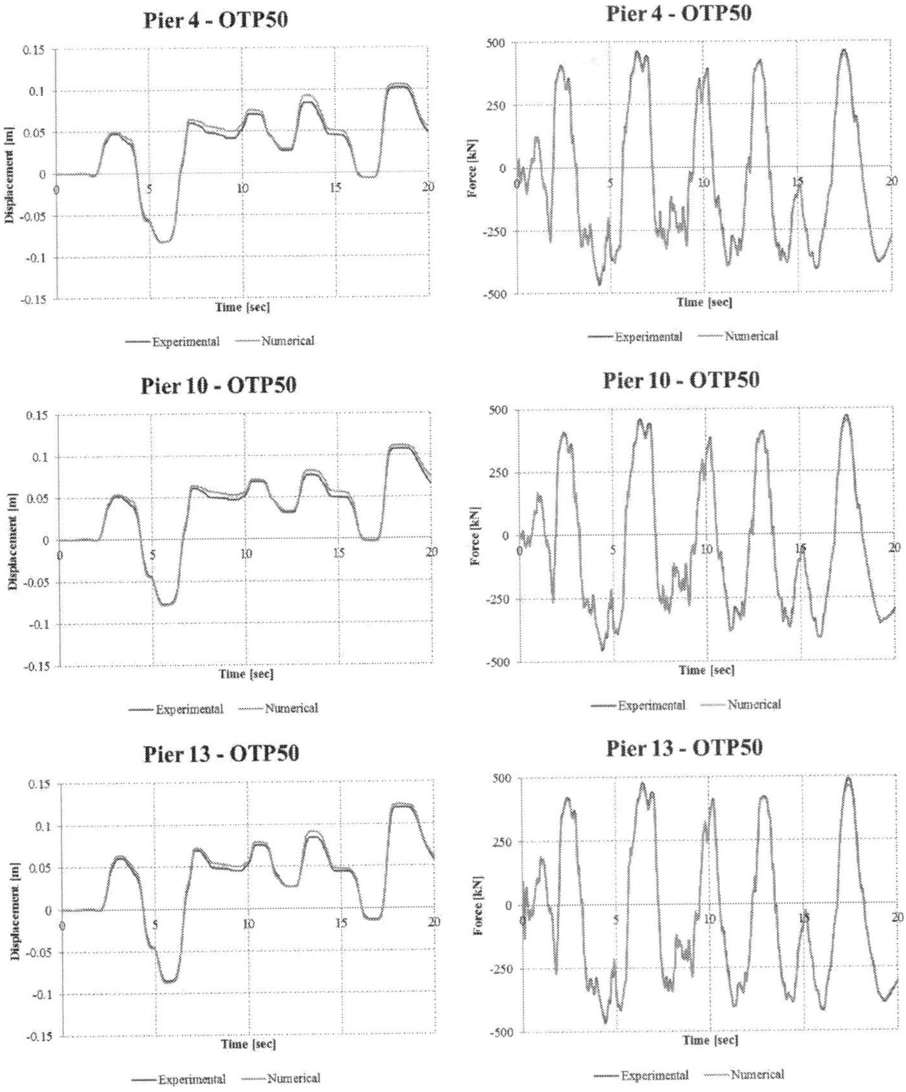

Figure 12. Charts displacement versus time and force versus time of the OTP and OP devices—comparison between experimental and numerical results.

	P1	P2	P3	P4	P5	P6	P7	P8	P9	P10	P11	P12	P13	P14	Tot	q
Non-linear analysis	772	939	1143	1483	1156	1076	1208	1192	1144	1466	1068	1092	1508	926	16173	1.5
Linear analysis	685	850	1438	2554	2274	1608	1437	1417	2580	2551	1560	2206	2482	900	24542	

Figure 13. Diagrams longitudinal moment curvature at the base of the piers—comparison between experimental and numerical results.

It is easy to find the substantial congruence of the two analyses in terms of forces and displacements, which in the diagrams in **Figure 12** are compared in a time history analysis.

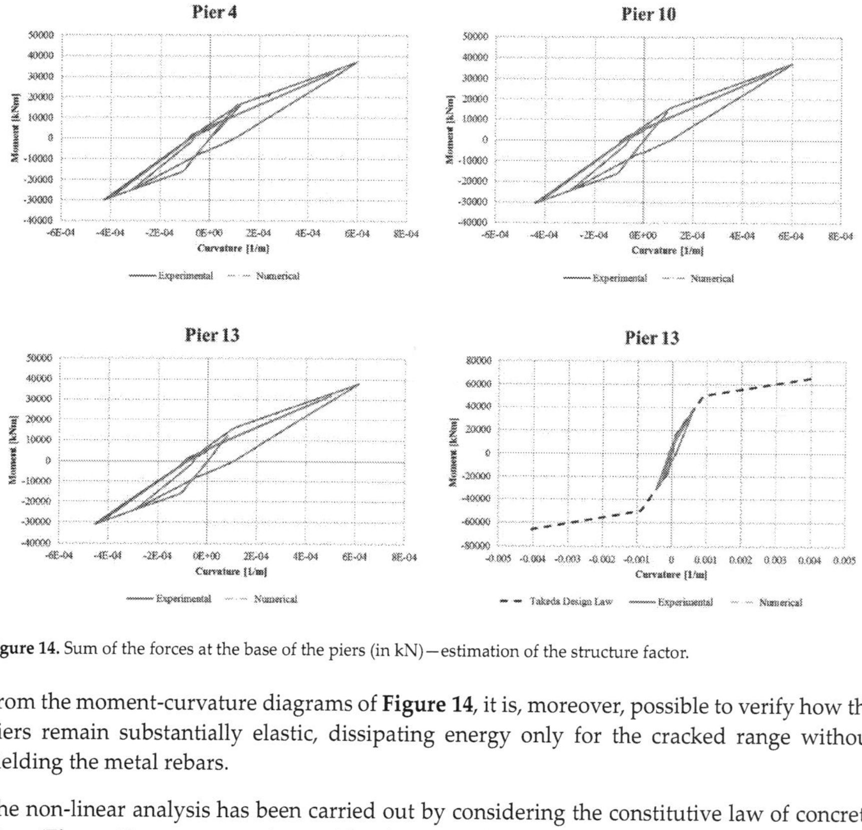

Figure 14. Sum of the forces at the base of the piers (in kN)—estimation of the structure factor.

From the moment-curvature diagrams of **Figure 14**, it is, moreover, possible to verify how the piers remain substantially elastic, dissipating energy only for the cracked range without yielding the metal rebars.

The non-linear analysis has been carried out by considering the constitutive law of concrete piers (**Figure 9**), so a comparison with a linear analysis is necessary in order to assess and confirm the structure factor, adopted in the verifications. The structure factor is equal to $q = 1.5$ and derives from the ratio of the summation of the shear force at the base of each pier (**Figure 13**).

The superstructure is requested to remain elastic during the seismic event. To control this requirement, the set of largest transversal displacements relating to the top of the piers has been applied to an accurate finite element model of the deck, where the girders and the slab are represented with shell elements, the bracing system are modelled with linear element and the devices are springs characterized by their stiffness.

The displacement set is the output of the non-linear analysis, with a peak of 0.33 m at the pier 7 (**Figure 15**).

It follows the calculation of stresses on the steel elements (**Figure 16**), where it is notable that the stress level is lower than the yielding limit.

Pier	Displacement [m]	Pier	Displacement [m]
P1	0.0008	P8	0.2198
P2	-0.0029	P9	0.0885
P3	-0.0177	P10	0.0813
P4	-0.0229	P11	0.1739
P5	0.0722	P12	0.1349
P6	0.2460	P13	0.0502
P7	0.3246	P14	0.001

Figure 15. Transversal displacement of the piers applied to the FEM model.

Figure 16. Stresses on the steel elements.

4. Conclusions

Construction methods and seismic protection techniques are interesting topics regarding multi-span viaducts characterized by steel-concrete composed deck.

A detailed design leads to the solution of many issues deriving from the structural altimetric and planimetric irregularity and from the high seismicity of the site.

The analyses have been performed applying the capacity design approach. Dissipative devices have been used against the seismic forces in order to grant and control that the piers remain substantially elastic.

For such purpose, each pier has been modelled taking into account its non-linear behaviour (theory of Takeda), simulating the stiffness at the cracking and plastic deformation status, in terms of moment curvature.

It has been verified that the system displacement request is lower than its capacity; the comparison between the ductility requested and the ductility available for the piers has been carried out according to the instructions provided by EC8-2 Annex E.

By the application of such design philosophy, it was possible to limit the forces at the base of the piers and, on the other side, the displacements of the decks and of the top of the same piers; the linear behaviour requirement of the metal superstructure under the effect of the seismic action was fulfilled as well.

At last, the comparison between the results of the FEM analysis performed using numerical and experimental data highlighted the correctness of the design hypothesis assumed.

Author details

Alessandro Contin and Andrea Mardegan*

*Address all correspondence to: andrea.mardegan@e2b.it

E2B s.r.l., Padova, Italy

References

[1] UNI EN 15129:2009. Anti-seismic devices.

[2] Nuove Norme Tecniche per le Costruzioni – DM Infrastrutture 14/01/2008, G.U. 4/2/2008 n.29 NTC (2008).

[3] Consiglio superiore dei lavori pubblici. Istruzioni per l'applicazione delle "Norme tecniche per le costruzioni" di cui al D.M. 14/01/2008.

[4] UNI EN 1998-2:2012. Eurocode 8: Design of structures for earthquake resistance. Part 2: bridges.

[5] Takeda T., Sozen M.A., Norby Nielsen N., Reinforced Concrete Response to Simulated Earthquakes, Journal of the Structural Division, Proceeding of the American Society of Civil Engineers, December 1970.

Chapter 7

Nondestructive Testing Structural Bridge Identification

Ahmet Can Altunışık

Additional information is available at the end of the chapter

http://dx.doi.org/10.5772/64288

Abstract

The bridges are one of the most important engineering structures. Determination of the bridge responses during their service life has gained great importance using nondestructive test methods with the changing of aims, usages, environmental conditions, material deteriorations by time, and damages during some dramatical events. This chapter presents the nondestructive experimental measurement test results of the bridges for structural identification. Ten different bridges, which have different type and carrier systems, such as historical masonry arch bridges, long span concrete highway bridges, base isolated bridges, footbridges, steel bridges, and old riveted bridges, are selected for numerical examples. The measurements are conducted under environmental excitations of pedestrian movement, traffic, wind-induced vibration, and the response signals are collected using uniaxial- and triaxial-sensitive seismic accelerometers. Operational modal analysis or ambient vibration tests are performed to extract the dynamic characteristics such as natural frequencies, mode shapes, and damping rations using enhanced frequency domain decomposition method in the frequency domain and stochastic subspace identification method in the time domain. It is demonstrated that the ambient vibration measurements are enough to identify the most significant modes of all bridge types.

Keywords: ambient vibration test, dynamic characteristics, nondestructive test methods, operational modal analysis

1. Introduction

The bridges, which link past to present and age gracefully, are one of the most important engineering structures. The bridges with different characteristics, thanks to their views, effects, and feelings during passing on, holds around and locations bring together the people for ages. In early applications, the bridges were designed as short span and narrow with stone and wood materials, and be able to carry light loads. But, nowadays, these conventional bridges have been replaced to steel and reinforced concrete.

There are various bridge types constructed during the last century according to the carrier system type, span lengths, and material properties such as masonry arch bridges, long span concrete/steel/composite highway bridges, base isolated bridges, footbridges, steel bridges, suspension bridges, cable-stayed bridges, and wooden/timber bridges. Masonry bridges have been built worldwide for social, economic, and strategic purposes. Originally intended to carry only pedestrian and horse-drawn vehicles, many of these historical bridges currently serve as critical components of transportation systems and, thus, must withstand significantly larger loads. Among various types of civil engineering structures, long span highway bridges, which are commonly used for passing large rivers, dam reservoirs, and deep valleys, attract the greatest interest for study particularly in terms of structural performance. Footbridges are generally situated to allow pedestrians to cross water or railways in areas where there are no nearby roads to necessitate a road bridge, and also across busy roads to let pedestrians cross safely without slowing down the traffic. Steel offers many advantages to the bridge builder, not only the material itself, but also its broad architectural possibilities such as high strength-to-weight ratio, high-quality material, speed of construction, versatility, modifications, recycling, durability, and aesthetics. Suspension and cable-stayed bridges are widely used across long spans (>550 m) and give rise to the usage of domains under the bridge. For this reason, the uses of suspension and cable-stayed bridges have increased recently. Wood is one of the most used and common materials for bridge constructions from the ancient times when humans first started finding ways on how to cross rivers and hard terrains.

Determination of dynamic response of bridges under static and dynamic loads, such as wind, earthquake, or traffic, is very complex and requires special studies. Finite element method has been widely used in civil engineering application since 1950s. Static, dynamic, linear, and nonlinear behavior can be obtained and illustrated using this method. It is generally expected that finite element models (FEMs) based on technical design data and engineering judgments can yield reliable simulation. However, because of modeling uncertainties, these models often cannot predict dynamic characteristics with the required level of accuracy. This raises the need for verification of finite element models using nondestructive experimental measurement tests.

There are two basically different methods available to experimentally identify the dynamic system parameters of a structure: experimental modal analysis (EMA) and operational modal analysis (OMA). In the EMA, the structure is excited by known input forces and the structural behavior is evaluated. In the OMA, the ambient vibrations such as vehicle load, wind, or wave loads have been used to actuate the structures. Heavy forced excitations may become expensive and sometimes may cause damage to the structure. Ambient excitations and their combination are environmental or natural excitations. Structural identification using this method gains the major importance. In this case, only response data of ambient vibrations are measurable while actual loading conditions are unknown. A system identification procedure will therefore need to base itself on output-only data.

It is well accepted that the finite element model updating is used to minimize the differences between analytically and experimentally determined dynamic characteristics by changing some uncertain parameters such as material properties, boundary conditions, section and connection details, and some additional structural elements and weights. In the finite element model updating, determination of the uncertain parameters and their ratios/values can be decided according to the nondestructive testing methods such as visual inspection, half-cell electrical potential method, Schmidt rebound hammer test, carbonation depth measurement test, permeability test, penetration resistance or Windsor probe test, resistivity measurement, electromagnetic methods, radiographic testing, ultrasonic testing, infrared thermography, ground penetrating radar, radioisotope gauges, acoustics emission, computed tomography, strain sensing, and corrosion rate measurement. The detailed information can be found in the related literature.

2. Modal parameter estimation methods

2.1. Enhanced frequency domain decomposition (EFDD) method

Enhanced frequency domain decomposition (EFDD) method is an extension of frequency domain decomposition (FDD) method which is a basic and easy-to-use method. In this method, modes are simply picked locating the peaks in singular value decomposition plots [1, 2].

In EFDD, the single degree of freedom (SDOF) power spectral density (PSD) function, identified around a peak of resonance, is taken back to the time domain using the Inverse Discrete Fourier Transform. In EFDD method, the relationship between unknown input and measured responses can be expressed as [2, 3]:

$$\left[G_{yy}(jw)\right]=\left[H\{jw\}\right]^{*}\left[G_{xx}\{jw\}\right]\left[H(jw)\right]^{T} \tag{1}$$

where $G_{xx}\{jw\}$ is the PSD matrix of the input, $G_{yy}\{jw\}$ is the PSD matrix of the responses, $H\{jw\}$ is the frequency response function (FRF) matrix, and * and superscript T denote complex conjugate and transpose, respectively. The FRF can be written in partial fraction, i.e., pole/residue form as [4]

$$H\{jw\}=\sum_{k=1}^{n}\frac{R_k}{jw-\lambda_k}+\frac{R_k^*}{jw-\lambda_k^*} \tag{2}$$

where n is the number of modes, λ_k is the pole, and, R_k is the residue. Substituting Eq. (2) into (1), we have

$$G_{yy}(jw)=\sum_{k=1}^{n}\sum_{s=1}^{n}\left[\frac{R_k}{jw-\lambda_k}+\frac{R_k^*}{jw-\lambda_k^*}\right]\left[G_{xx}\{jw\}\right]\left[\frac{R_s}{jw-\lambda_s}+\frac{R_s^*}{jw-\lambda_s^*}\right]^{H} \tag{3}$$

where s is the singular value, superscript H denotes complex conjugate and transpose. Multiplying the two partial fraction factors and making use of the Heaviside partial fraction theorem, the output PSD can be reduced to a pole/residue form as follows:

$$G_{yy}(jw) = \sum_{k=1}^{n} \frac{A_k}{jw - \lambda_k} + \frac{A_k^*}{jw - \lambda_k^*} + \frac{B_k}{-jw - \lambda_k} + \frac{B_k^*}{-jw - \lambda_k^*} \tag{4}$$

where A_k is the kth residue matrix of the output PSD. In the EFDD identification, the first step is to estimate the PSD matrix. The estimation of the output PSD, $G_{yy}(jw)$ known at discrete frequencies $w = w_i$ is then decomposed by taking the SVD of the matrix

$$G_{yy}(jw_i) = U_i S_i U_i^H \tag{5}$$

where the matrix $U_i = [u_{i1}, u_{i2}, \dots, u_{im}]$ is a unitary matrix holding the singular vectors, u_{ij}, and S_i is a diagonal matrix holding the scalar singular values s_{ij} [4–6].

2.2. Stochastic subspace identification (SSI) method

Stochastic subspace identification (SSI) method is an output-only time domain method that directly works with time data, without the need to convert them to correlations. The model of structural vibrations can be defined by a set of linear, constant coefficient and second-order differential equations [7]:

$$M\ddot{U}(t) + C_2\dot{U}(t) + KU(t) = F(t) = B_2 u(t) \tag{6}$$

where M, C_2, and K are the mass, damping, and stiffness matrices, $F(t)$ is the excitation force, and $U(t)$ is the displacement vector depending on time t. Note that the force vector $F(t)$ is factorized into a matrix B_2 describing the inputs in space and a vector $u(t)$. The equation of dynamic equilibrium (6) will be converted to a more suitable form: the discrete-time stochastic state-space model [7, 8]. With the following definitions

$$x(t) = \begin{pmatrix} U(t) \\ \dot{U}(t) \end{pmatrix}, \; A = \begin{pmatrix} 0 & I_{n_2} \\ -M^{-1}K & -M^{-1}C_2 \end{pmatrix}, \; B = \begin{pmatrix} 0 \\ M^{-1}B_2 \end{pmatrix} \tag{7}$$

Eq. (6) can be transformed into the state equation

$$\dot{x}(t) = Ax(t) + Bu(t) \tag{8}$$

where A is the state matrix, B is the input matrix, and $x(t)$ is the state vector. If it is assumed that the measurements are evaluated at only one sensor location, and that this sensor can be accelerometer, velocity, or displacement transducer, the observation equation is [9]:

$$y(t) = Cx(t) + Du(t) \tag{9}$$

where C is the output matrix and D is the direct transmission matrix. Eqs. (8) and (9) constitute a continuous-time deterministic state-space model. This is not realistic: measurements are available at discrete time instants $k\Delta t$, $k \in N$ with Δt, the sample time and noise is always influencing the data. After sampling, the state-space model looks like [6]:

$$\begin{aligned} x_{k+1} &= Ax_k + Bu_k \\ y_k &= Cx_k + Du_k \end{aligned} \tag{10}$$

where $x_k = x(k\Delta t)$ is the discrete-time state vector. The stochastic components are included and obtained discrete-time combined deterministic-stochastic state-space model:

$$\begin{aligned} x_{k+1} &= Ax_k + Bu_k + w_k \\ y_k &= Cx_k + Du_k + v_k \end{aligned} \tag{11}$$

where w_k is the process noise due to disturbances and modeling inaccuracies and v_k is the measurement noise due to the sensor inaccuracy. They are both immeasurable vector signals but we assume that they are zero mean, white, and covariance matrices [7]:

$$E\left[\begin{pmatrix} w_p \\ v_p \end{pmatrix}\begin{pmatrix} w_q^T & v_q^T \end{pmatrix}\right] = \begin{pmatrix} Q & S \\ S^T & R \end{pmatrix}\delta_{pq} \tag{12}$$

where E is the expected value operator and δ_{pq} is the Kronecker delta. This is a function of two variables, usually integers, which is 1 if they are equal, and 0 otherwise. It can be written as the symbol δ_{pq}, and treated as a notational shorthand rather than a function:

$$\delta_{pq} = \begin{cases} 1, & if\ p = q \\ 0, & if\ p \neq q \end{cases} \tag{13}$$

The vibration information that is available in structural health monitoring is usually the responses of a structure excited by the operational inputs that are some immeasurable inputs.

It is impossible to distinguish deterministic input u_k from the noise terms w_k, v_k in Eq. (11). If the deterministic input term u_k is modeled by the noise terms w_k, v_k the discrete-time purely stochastic state-space model is obtained:

$$\begin{aligned} x_{k+1} &= Ax_k + w_k \\ y_k &= Cx_k + v_k \end{aligned} \tag{14}$$

Eq. (14) constitutes the basis for the time-domain system identification through operational vibration measurements.

2.3. Modal assurance criterion

The modal assurance criterion (MAC) is defined as a scalar constant relating the degree of consistency (linearity) between one modal and another reference modal vector [10] as follows:

$$MAC = \frac{\left|\{\varnothing_{ai}\}^T\{\varnothing_{ej}\}\right|^2}{\{\varnothing_{ai}\}^T\{\varnothing_{ai}\}\{\varnothing_{ej}\}^T\{\varnothing_{ej}\}} \tag{15}$$

where $\{\varnothing_{ai}\}$ and $\{\varnothing_{ej}\}$ are the modal vectors of ith and jth for different methods, respectively.

3. Nondestructive testing of bridges

In the content of this chapter, 10 different bridges that have different type and carrier systems are selected as case studies:

- Historical masonry arch bridges (Osmanlı, Mikron, and Şenyuva)
- Long span concrete highway bridges (Kömürhan and Birecik)
- Base isolated bridge (Gülburnu)
- Footbridges (Ortahisar and Akçaabat)
- Steel bridges (Eynel)
- Old riveted bridges (Borçka)

3.1. Historical masonry arch bridges

Historical structures are identity of the communities. They are not only structures, which contain stone, timber, mortar, etc., they also contain the social culture and this is the biggest difference between the new structures. Almost every person is curious about the past and they want to learn some information of their ancestors. So the easiest way to learn about the past is to examine the historical data and structures. In the last century, people have given more attention to preserve the historical structures. A lot of studies have been carried out for

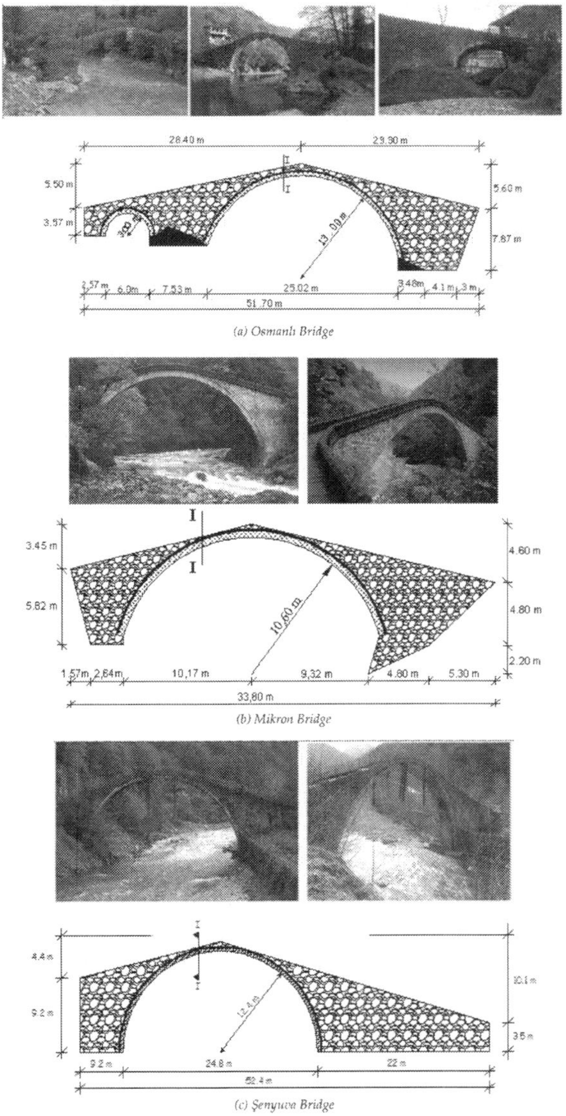

Figure 1. Views of the historical masonry arch bridges with relieve drawings. (a) Osmanlı Bridge, (b) Mikron Bridge, and (c) Şenyuva Bridge.

estimating behavior of these structures and reliable restoration could be made to preserve them for future.

Masonry arch bridges hold an important place in historical structures [11]. They are not complex structures. A stone arch bridge consists of stone blocks and mortar joints. Blocks have high strength in compression and low strength in tension while mortar has generally low strength. Historical masonry arch bridges are vital components of transportation systems in many countries worldwide, ensuring the ready access of goods and services to millions of people [12]. Many of those bridges, which were originally built for the passage of carts, are being used for road and rail vehicles. They demonstrate a surprisingly high load bearing capacity and good durability.

Osmanlı, Mikron, and Şenyuva historical masonry arch bridges constructed in Turkey are selected for example. The Osmanlı historical masonry arch bridge was built in the nineteenth

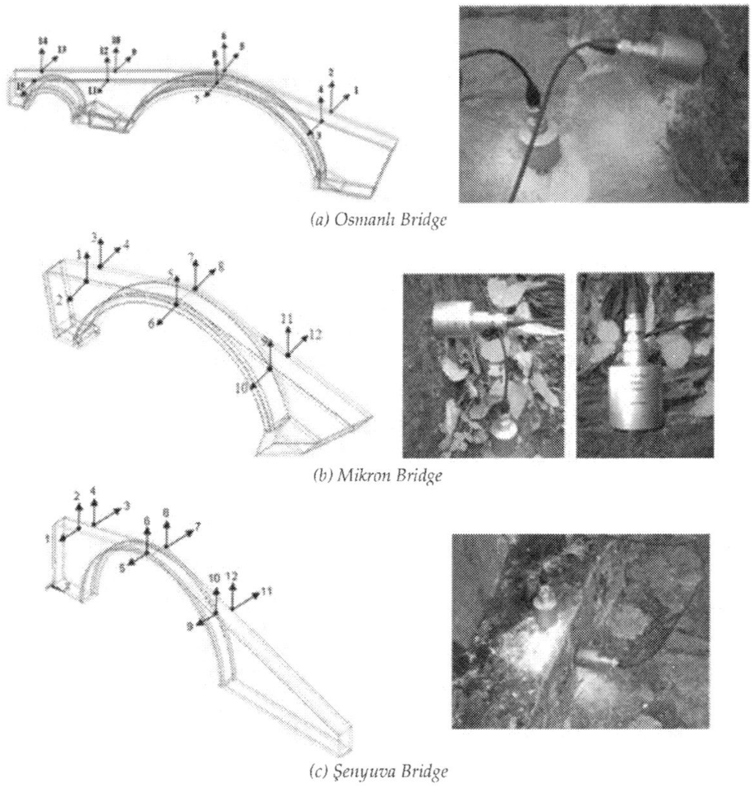

Figure 2. Accelerometer locations and views from the measurements. (a) Osmanlı Bridge, (b) Mikron Bridge, and (c) Şenyuva Bridge.

century. This two-spanned arch bridge has a total length of 51.7 m. The span of each arch is 25.2 and 6 m, and the radius of each arch is 13 and 3m, respectively.

The Mikron historic arch bridge, built in the mid-nineteenth century during the Ottoman Empire, spans the Fırtına River in Rize, Turkey. Cut stone blocks composed the bridge's arches and parapets. In 1998, Turkey's General Directorate for Highways supervised repair of the main structural elements of the bridge (stone arches, side walls, and filler material). The bridge, with a total length of 33.80 m, has two stone, inner and outer semi-circular arches, which have thicknesses of 0.50 and 0.15 m, respectively.

The Şenyuva historical arch bridge built in 1696 by the native population is located on Fırtına Stream in Çamlihemşin, Rize, Turkey. The bridge has a single arch. The total span of bridge is 52.4 m, the span of the bridge arch is 24.8 m, the height of the arch is 12.4 m, and the wide of the deck is 2.5 m. Height of the side walls at both side are 9.2 and 3.5 m, respectively. There are 60 cm × 30 cm dimensional parapets on both sides of the bridge deck. Some views of the bridges with relieve drawings are given in **Figure 1(a–c)**.

Ambient vibration tests were performed under existing environmental condition. B&K 8340 and B&K 3560 experimental measurement equipment were used. PULSE [13] and OMA [14] softwares were used to signal processing and parameter estimations. Accelerometer locations and views from the measurements for each bridge are shown in **Figure 2**.

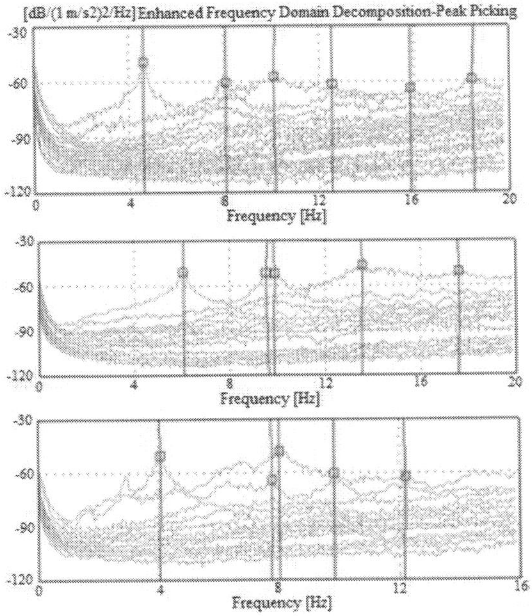

Figure 3. The singular values of spectral density matrices for historical masonry arch bridges.

Accelerometer setups shown in **Figure 2** were used, and measurements were carried out for at least 30 min. The singular values of spectral density matrices are given in **Figure 3**. The dynamic characteristics are given in **Table 1** and **Figure 4**. The first four natural frequencies are obtained between 4 and 14 Hz. The mode shapes occurred as lateral and vertical forms.

Bridges	**Natural frequencies (Hz)**				**Damping ratio (%)**			
	1	**2**	**3**	**4**	**1**	**2**	**3**	**4**
	EFDD method							
Osmanlı	4.640	8.094	9.879	12.340	1.634	1.035	6.157	0.256
Mikron	6.063	9.563	9.906	13.590	1.945	0.967	0.835	0.258
Şenyuva	4.045	7.750	8.020	10.000	2.377	1.318	4.288	0.265
	SSI method							
Osmanlı	4.642	8.325	9.735	11.910	1.634	1.035	6.157	0.256
Mikron	6.065	9.558	10.180	13.590	1.855	0.923	0.815	0.289
Şenyuva	4.066	7.960	8.044	10.100	2.377	1.318	4.288	0.265

Table 1. Experimentally identified natural frequencies and damping ratios.

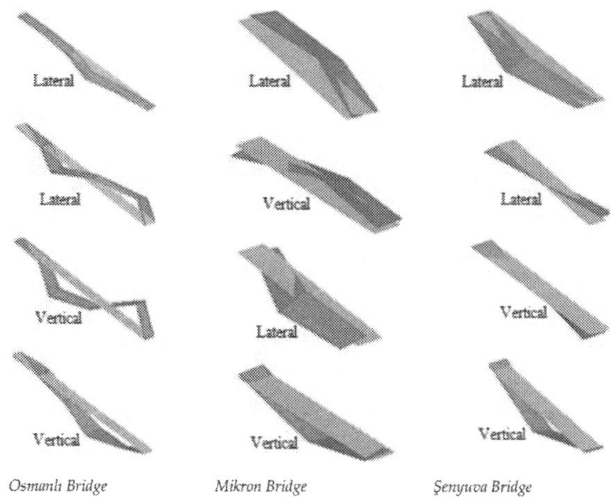

Figure 4. The first four mode shapes of the historical masonry arch bridges.

3.2. Long span concrete highway bridges

Kömürhan and Birecik long span concrete highway bridges constructed in Turkey are selected for example. The bridge deck consists of a main span of 135 m and two side span of 76 m each.

The total bridge length is 287 m and width of the bridge is 11.50 m. The structural system of Kömürhan Highway Bridge consists of deck, columns, side support, and expansion joint. The deck of the bridge was constructed with balanced cantilever and prestressed box beam method. There are two main columns of 59.50 m each. Foundation of the main column is concrete in mass having the dimension of 24 m × 13.5 m and 5 m depth. To combine deck cantilevers, an expansion joint is constituted in the main span of the bridge [16, 17].

(a) Kömürhan Bridge

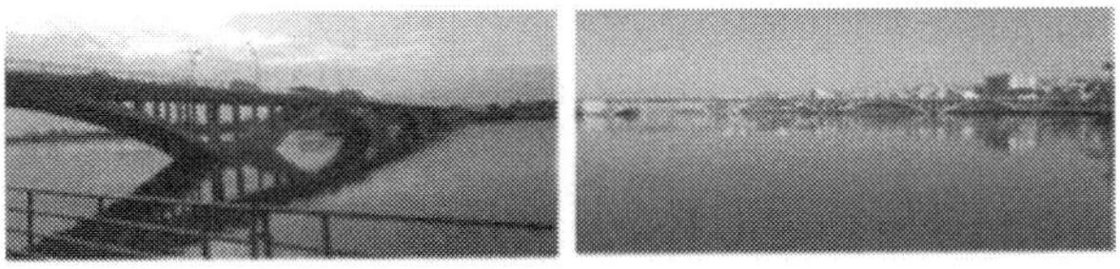

(b) Birecik Bridge

Figure 5. Some views of the long span concrete highway bridges with relieve drawings. (a) Kömürhan Bridge and (b) Birecik Bridge.

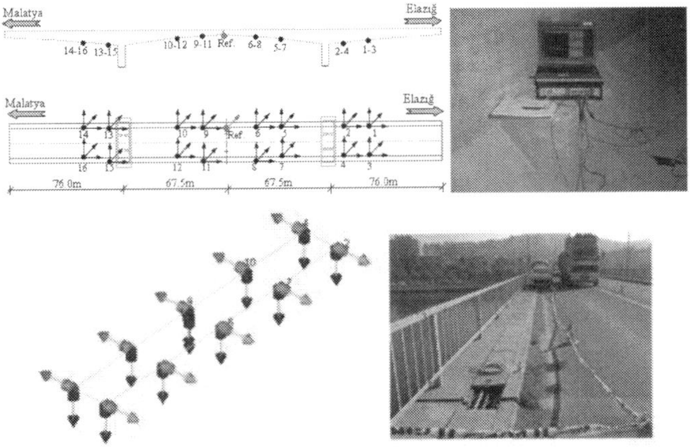

Figure 6. Accelerometer location and views from the measurements.

Birecik Bridge is located 81 km of the Şanlıurfa-Gaziantep state highway over Fırat River in Turkey. The construction of the bridge was started in June 1951 and the bridge was opened to the traffic in April 1956. The bridge consist of five arches, each arch has a 55 m main span. The total bridge length is 300 m and the width of the bridge is 10 m. The bridge arches have rigid connectivity at middle spans and side supports. But, right and left side of the middle points of slabs are constructed using joints. Columns, beams, arches, decks, and foundations were constructed as reinforced concrete [18]. Some views of the bridges with relieve drawings is given in **Figure 5**. **Figure 6** presents the accelerometer location and views from the measurement. The measurements were carried out for at least 60 min. The singular values of spectral density matrices obtained from vibration data are given in **Figure 7**.

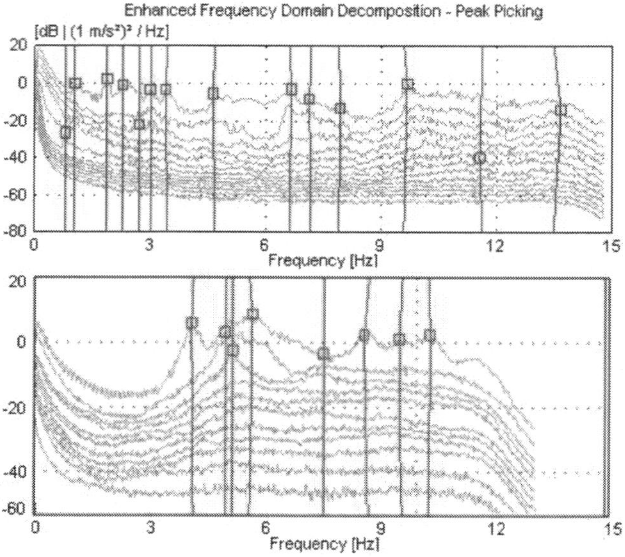

Figure 7. The singular values of spectral density matrices for long span highway bridges.

The dynamic characteristics and related mode shapes are given in **Table 2** and **Figure 8**. The first four natural frequencies are obtained between 0.7 and 2.3 Hz for the Kömürhan Bridge and 2.4 and 4.6 Hz for the Birecik Bridge, respectively. The mode shapes occurred in lateral, vertical, and torsional forms.

Bridges	**EFDD method**							
	Natural frequencies (Hz)				**Damping ratio (%)**			
	1	**2**	**3**	**4**	**1**	**2**	**3**	**4**
Kömürhan	0.788	1.027	1.850	2.291	1.373	1.785	2.057	1.465
Birecik	2.496	3.115	3.378	4.545	4.358	0.899	0.863	0.118

Table 2. Experimentally identified natural frequencies and damping ratios.

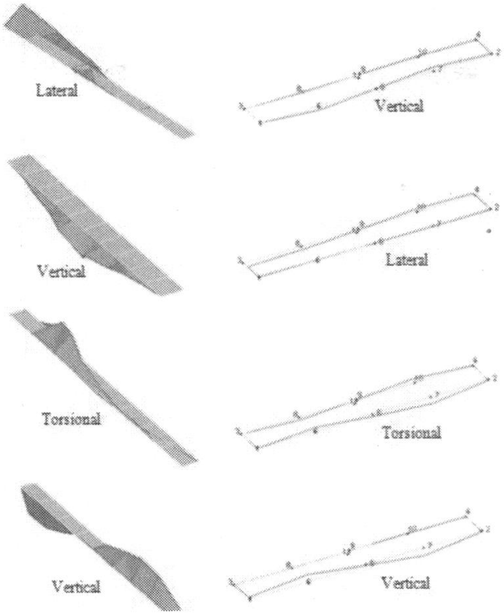

Figure 8. The first four mode shapes of the long span highway bridges.

3.3. Base isolated bridge

Base isolated Gülburnu Highway Bridge constructed in Turkey is selected for example. The construction of the bridge was started in November 2005 and the bridge was opened to the traffic in May 2009. The bridge is twin prestressed concrete box girder structures. The bridge

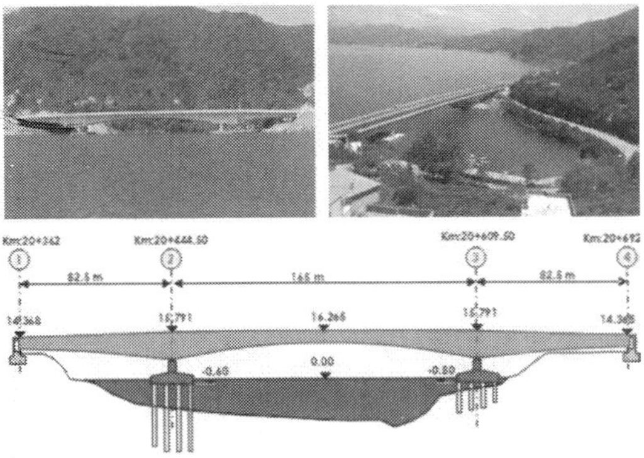

Figure 9. Some views of the base isolated bridge with relieve drawings.

deck consists of a main span of 165 m and two side span of 82.5 m each. The total bridge length is 330 m and the width of the bridge is 30 m. The structural system of the bridge consists of deck, piers, and side support. There are four piers and each has 4.50 m height and 9.00 m × 3.75 m cross-section areas. All piers are footed on the two raft foundation with bored piles. Two abutments that allow longitudinal direction movement only support the superstructure at both sides [19, 20]. Some views of the bridge with relieve drawings is given in **Figure 9**.

Accelerometer locations are shown in **Figure 10**. The measurements were carried out for at least 60 min. The singular values of spectral density matrices obtained from the processing vibration data are given in **Figure 11**.

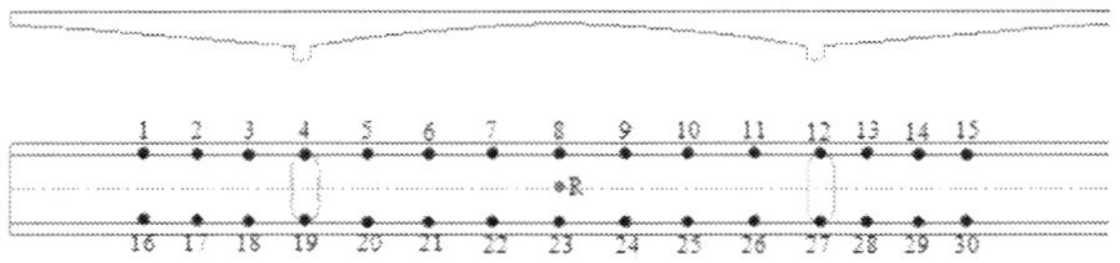

Figure 10. Accelerometer location and views taken from the measurements.

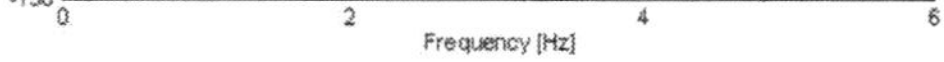

Figure 11. The singular values of spectral density matrices for base isolated bridge.

Mode	Natural frequencies (Hz)		Damping ratio (%)	
	EFDD	SSI	EFDD	SSI
1	0.993	0.995	2.661	3.952
2	1.508	1.505	0.958	0.559
3	2.238	2.241	0.741	0.604
4	2.853	2.874	0.765	2.145
5	3.181	3.258	0.371	0.758
6	4.321	4.298	0.558	0.962

Table 3. Experimentally identified natural frequencies and damping ratios.

Natural frequencies, mode shapes, and damping ratios are given in **Table 3** and **Figure 12**. The first six natural frequencies are obtained between 0.9 and 4.5 Hz. The mode shapes occurred in vertical, torsional, longitudinal, and lateral forms.

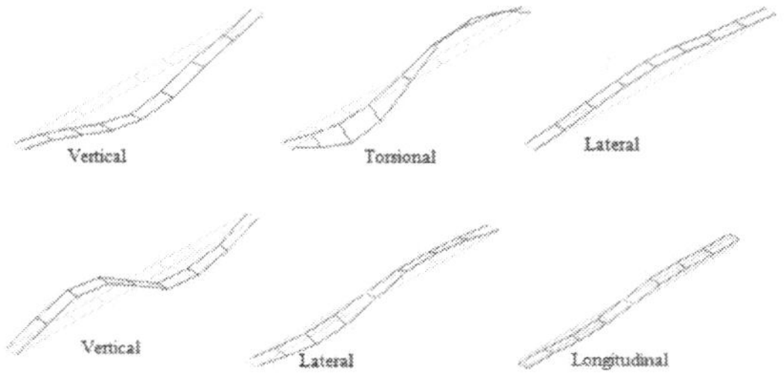

Figure 12. The first six mode shapes of the base isolated bridge.

3.4. Footbridges

Ortahisar and Akçaabat footbridges constructed in Turkey are selected for example. Ortahisar arch-type footbridge is located in a heavy traffic area in Trabzon, Turkey, and has a main span of 35 m. The footbridge operates as part of a pedestrian public footpath [21]. Akçaabat footbridge also operates as part of a pedestrian public footpath [22]. Some views of the footbridges with relieve drawings are given in **Figure 13**.

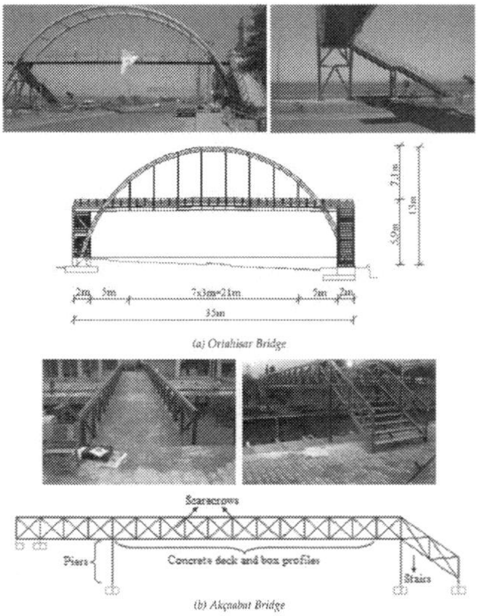

Figure 13. Some views of the footbridges with relieve drawings. (a) Ortahisar Bridge and (b) Akçaabat Bridge.

Some views from the measurements with accelerometer locations are shown in **Figure 14**. The measurements were carried out for at least 45 min. The singular values of spectral density matrices obtained from the processing vibration data are given in **Figure 15**.

The dynamic characteristics such as natural frequencies, mode shapes, and damping ratios obtained using EFDD and SSI methods in frequency and time domain are given in **Table 4** and **Figure 16**. The first six natural frequencies are obtained between 1.9 and 6.7 Hz. The mode shapes occurred in lateral and torsional forms.

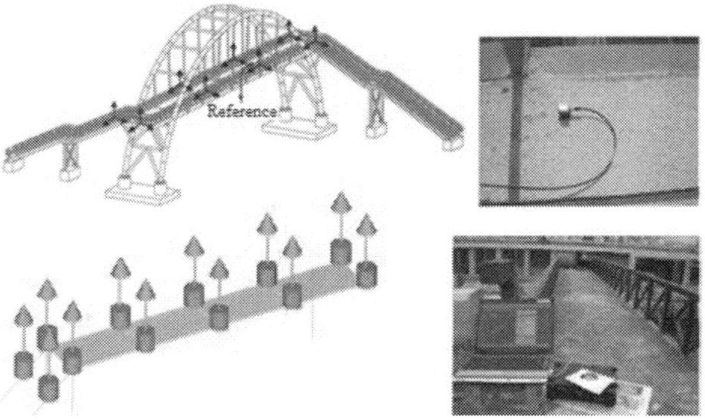

Figure 14. Accelerometer location and views taken from the measurements.

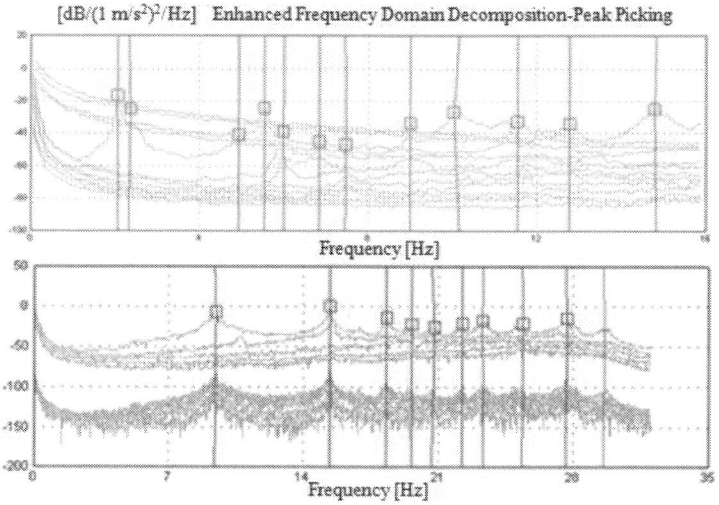

Figure 15. The singular values of spectral density matrices for footbridges.

http://dx.doi.org/10.5772/64288

Mode	Natural frequencies (Hz)		Damping ratio (%)
	EFDD	SSI	
1	2.08	1.90	1.22
2	2.34	2.40	2.82
3	4.78	4.70	0.37
4	5.53	5.50	0.84
5	6.01	5.80	0.40
6	6.67	6.70	0.26

Table 4. Experimentally identified natural frequencies and damping ratios.

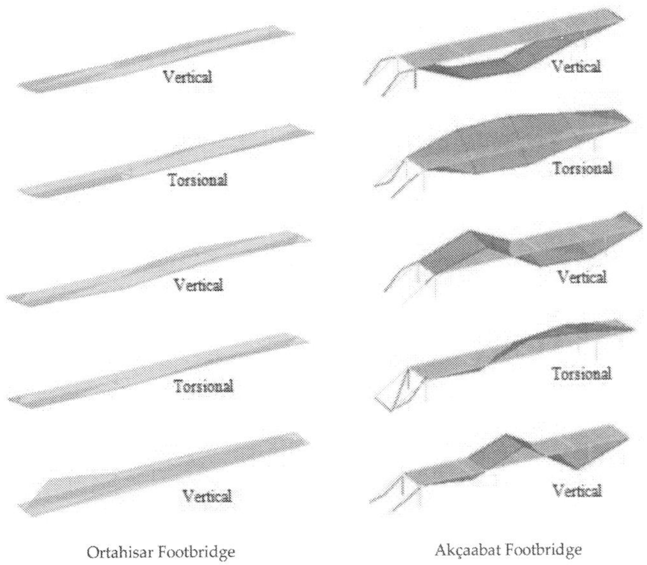

Figure 16. The first five mode shapes of the footbridges.

3.5. Steel bridges

Bridges are one of the most important engineering structures which are commonly used for interplant and intercity transportation. In Turkey, in earlier days they were designed as narrow and short span with stone and wood materials and to be able to carry light loads. Today, the location of these bridges has been replaced with long span reinforced concrete and steel bridges. According to the General Directorate of Highways data, there are 6447 highway bridges with a total length of 296 km in Turkey.

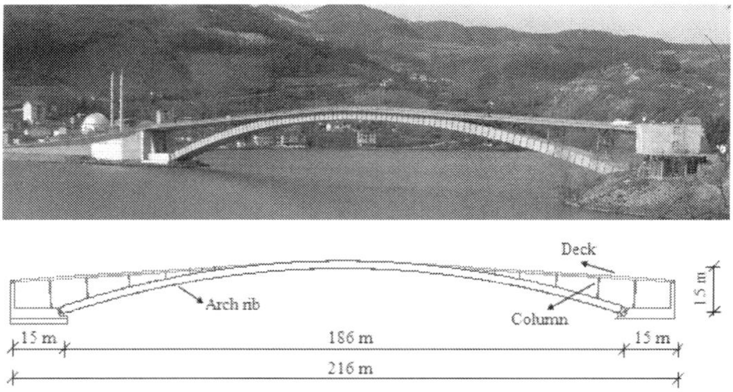

Figure 17. Some views of the steel bridge with relieve drawings.

Figure 18. Accelerometer location and views taken from the measurements.

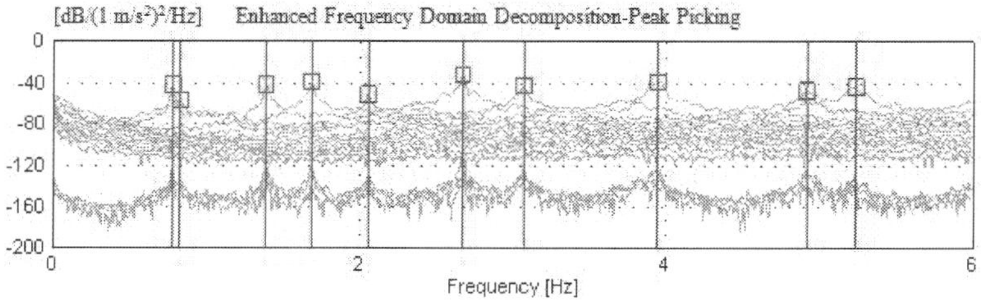

Figure 19. The singular values of spectral density matrices for steel bridge.

Eynel steel bridge constructed in Turkey is selected for example. The bridge is located in the Black Sea region of Turkey. It connects to the villages near the two sides of Suat Uğurlu Dam reservoir in the city of Samsun. The construction of the bridge started in 2007 and it was opened to traffic in 2009. The bridge is upper-deck steel bridge which has arch-type carriage system with a total length of 216 m. The span of the arch rib is 186 m and it has box-type section. The height and width of the section is 2.4 and 12 m. The deck is 12 m wide and has a constant thickness of 10 cm [23]. Some views of the steel bridge are given in **Figure 17**.

In **Figure 18**, the accelerometer locations and views during the measurement are presented in detail. The measurements were carried out for at least 60 min. The singular values of spectral density matrices obtained from the processing vibration data are given in **Figure 19**.

Table 5 and **Figure 20** summarize the dynamic characteristics obtained using EFDD and SSI methods. The first six natural frequencies are obtained between 0.7 and 2.7 Hz. The mode shapes occurred in lateral and transverse forms.

Mode	Natural frequencies (Hz)		Damping ratio (%)	
	EFDD	SSI	EFDD	SSI
1	0.779	0.800	0.73	1.67
2	0.828	-	1.30	-
3	1.395	1.381	0.51	1.06
4	1.688	1.709	0.40	3.26
5	2.057	1.933	0.45	0.84
6	2.674	2.670	0.25	0.36

Table 5. Experimentally identified natural frequencies and damping ratios.

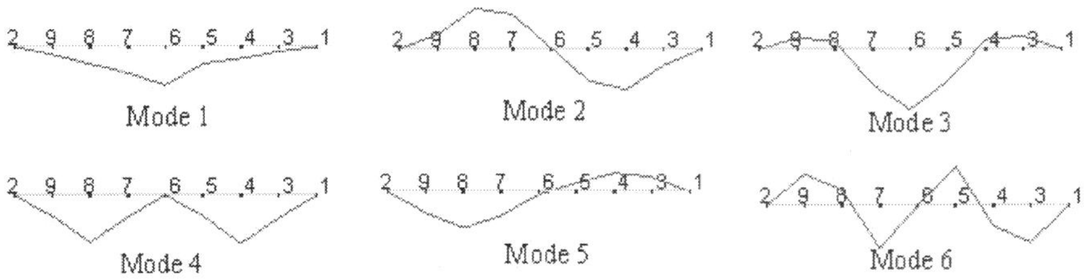

Figure 20. The first six mode shapes of the steel bridge.

3.6. Old riveted bridge

Borçka Old Riveted Bridge constructed in Turkey is selected for example. The bridge, built in 1936, is on the Çoruh River in the town center of Borçka. Total length and width of the bridge are about 114 and 5.30 m, respectively. The main structural system of the bridge has an arch height of 16.30 m from the bridge deck. The bridge girders consist of two edge beams and five middle beams in the longitudinal and transverse directions. The structural elements (arches, pillars, decks, wind connections, etc.) are made out of steel with riveted connections. Bridge is closed to vehicle traffic and it is for only pedestrians [24]. Some views of the old riveted bridge with relieve drawings are given in **Figure 21**.

Figure 21. Some views of the old riveted bridge with relieve drawings.

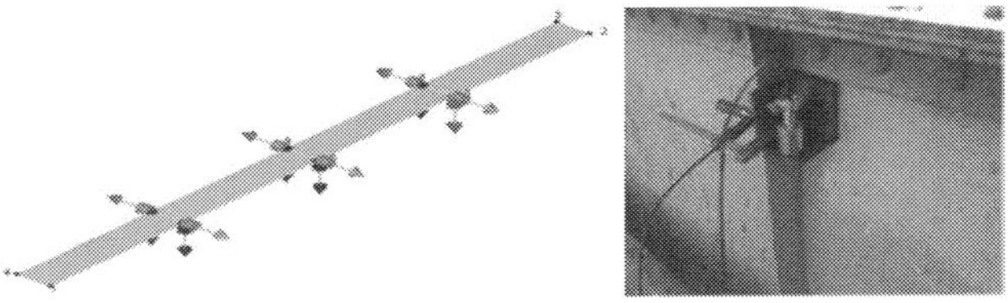

Figure 22. Accelerometer location and views taken from the measurement.

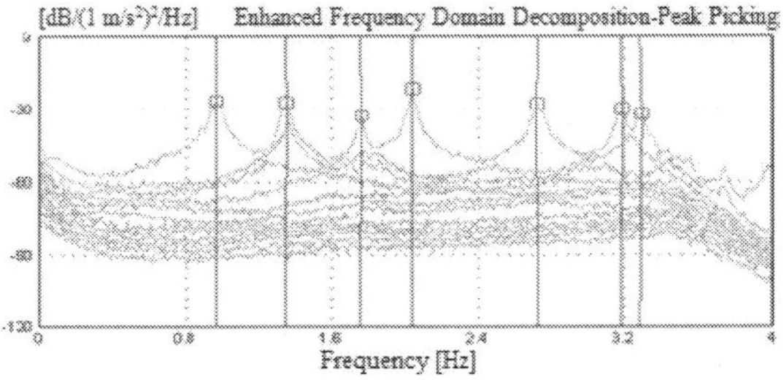

Figure 23. The singular values of spectral density matrices for old riveted bridge.

Mode	Natural frequencies (Hz)		Damping ratio (%)	
	EFDD	SSI	EFDD	SSI
1	0.970	0.968	2.185	1.801
2	1.352	1.348	0.736	0.926
3	1.761	1.758	0.962	0.817
4	2.042	2.041	0.459	0.401
5	2.726	2.725	0.764	0.707
6	3.183	3.189	0.432	0.395

Table 6. Experimentally identified natural frequencies and damping ratios.

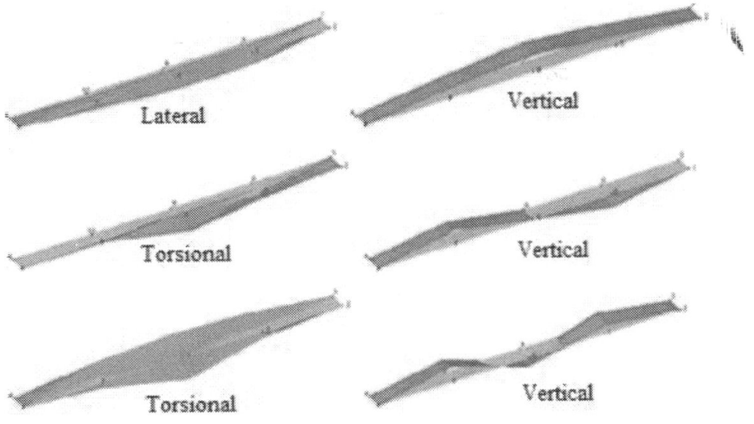

Figure 24. The first six mode shapes of the old riveted bridge.

The accelerometer locations and placements on the deck are displayed in **Figure 22**. The measurements were carried out for at least 60 min. The singular values of spectral density matrices obtained from the vibration data are given in **Figure 23**.

The dynamic characteristics obtained using EFDD and SSI methods are given in **Table 6** and **Figure 24**. The first six natural frequencies are obtained between 0.9 and 3.2 Hz. The mode shapes occurred in lateral, vertical, and torsional forms.

4. Finite element analyses and model updating

To validate the experimentally identified dynamic characteristics, finite element models of the bridges are constituted in the SAP2000 and ANSYS [25, 26] software. With modal analyses, the natural frequencies and related mode shapes are obtained (**Figure 25**). The analytically identified natural frequencies are summarized in **Table 7**.

It is seen that ambient vibration measurements are enough to identify the most significant modes of all bridge types. There is a good agreement between natural frequencies and corresponding mode shapes. The maximum differences are obtained as nearly as 10–20%. To eliminate these differences, the finite element models of the bridges should be updated by changing some uncertain parameters such as material properties, boundary conditions, section areas, etc. It can be evaluated that the maximum differences are reduced from 10–20 to 2–5% after model updating. More information can be found in the literature [11, 12, 15–18, 20–24, 27].

Also, to display the model updating effect, dynamic responses of the bridge are performed before and after model updating. It is seen that this procedure has vital importance to represent the real structural behavior. More information can be found in the literature [11, 12, 15–18, 20–24, 27].

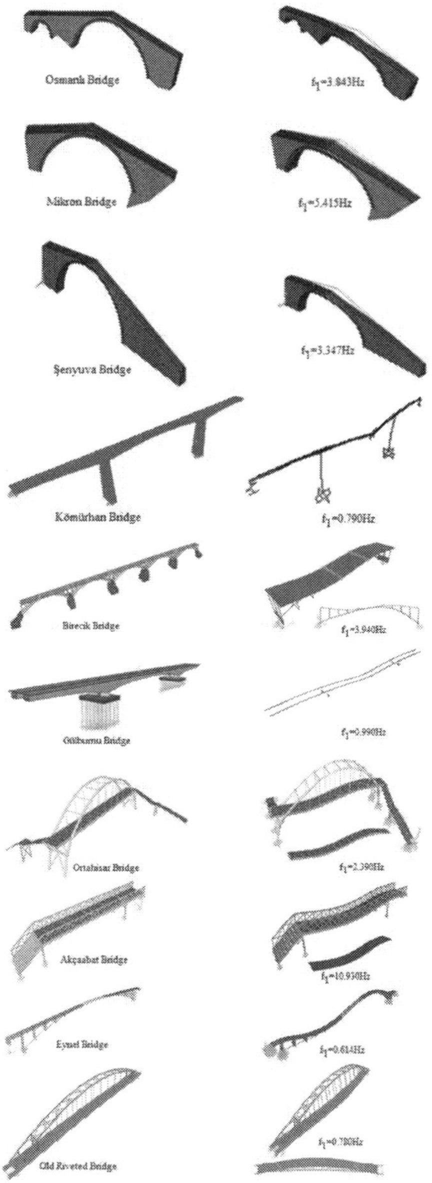

Figure 25. The finite element models of the bridges with first mode shapes.

Bridges		Natural Frequencies					
		1	2	3	4	5	6
Historical masonry	Osmanlı	3.843	7.527	9.371	10.638	14.563	–
	Mikron	5.415	10.113	10.665	13.371	–	–
	Şenyuva	3.347	5.772	7.754	9.055	10.044	–
Long span highway	Kömürhan	0.790	1.106	1.845	2.315	2.685	3.346
	Birecik	3.940	4.770	5.190	8.920	9.530	–
Base isolated	Gülburnu	0.990	1.485	2.164	2.742	3.150	4.186
Footbridge	Ortahisar	2.390	2.500	5.520	5.930	7.630	7.650
	Akçaabat	10.930	15.890	20.810	24.100	33.830	–
Steel	Eynel	0.614	0.718	1.186	1.754	1.940	2.386
Old riveted	Borçka	0.780	1.863	1.960	2.060	2.122	2.930

Table 7. Natural frequencies identified by finite element analyses.

5. Conclusion

This chapter presents a comparative study about the nondestructive measurement of bridges for structural identification. Ten different bridges, which have different type and carrier systems, are selected as case studies. The dynamic characteristics such as natural frequencies, mode shapes, and damping ratios are extracted using ambient vibration tests with operational modal analysis procedure. The experimentally identified dynamic characteristics are validated by the finite element results, and the differences are evaluated.

It can be seen that the ambient vibration measurements are enough to identify the most significant modes of all bridge types.

The first natural frequencies are obtained as 4–14, 0.7–2.3, 2.4–4.6, 0.9–4.5, 1.9–6.7, 0.7–2.7, and 0.9–3.2 Hz for historical masonry arch bridges, Kömürhan and Birecik long span highway bridges, base isolated bridge, footbridges, steel bridges, and old riveted bridges, respectively.

The mode shapes occurred as vertical, lateral, longitudinal, and torsional forms. Especially, longitudinal modes should be considered for a base isolated bridge.

The finite element analyses are performed, and the results are compared with each other. It is seen that there is a good agreement between the natural frequencies and corresponding mode shapes. The maximum differences are nearly within 10–20%.

To eliminate these differences, the finite element models of the bridges should be updated by changing some uncertain parameters such as material properties, boundary conditions, section areas, etc. It can be evaluated that the maximum differences are reduced from 10–20 to 2–5% after model updating procedure. More information can be found in the cited articles.

Acknowledgements

The results presented in this chapter are selected from the author and his co-authors' previous published works. The author was grateful to his advisor Prof. Alemdar Bayraktar and co-authors Assoc. Prof. Barış Sevim and Assist. Prof. Temel Türker for their valuable contributions.

Author details

Ahmet Can Altunışık

Address all correspondence to: ahmetcan8284@hotmail.com

Department of Civil Engineering, Karadeniz Technical University, Trabzon, Turkey

References

[1] Ren WX, Zhao T, Harik IE. Experimental and analytical modal analysis of steel arch bridge. Journal of Structural Engineering. 2004;130(7):1022–1031. DOI: 10.1061/(ASCE)0733-9445.

[2] Jacobsen NJ, Andersen P, Brincker R. Using enhanced frequency domain decomposition as a robust technique to harmonic excitation in operational modal analysis. In: Proceedings of International Conference on Noise and Vibration Engineering (ISMA2006); Leuven, Belgium; 2006.

[3] Bendat JS, Piersol AG. Random Data: Analysis and Measurement Procedures. John Wiley and Sons; USA; 2004.

[4] Brincker R, Zhang L, Andersen P. Modal identification from ambient responses using frequency domain decomposition. In: Proceedings of the 18th International Modal Analysis Conference, San Antonio, USA; 2000.

[5] Felber AJ. Development of hybrid bridge evaluation system. PhD Thesis. Vancouver, Canada: University of British Columbia; 1993.

[6] Peeters B. System identification and damage detection in civil engineering. PhD Thesis. Leuven, Belgium: K.U; 2000.

[7] Peeters B, DeRoeck G. Reference based stochastic subspace identification in civil engineering. In: Proceedings of the 2nd International Conference on Identification in Engineering Systems; Swansea, UK; 1999. Pp. 639–648.

[8] Ewins DJ. Modal Testing: Theory and Practice. England: Research Studies Press Ltd; 1984.

[9] Juang JN. Applied System Identification. Englewood Cliffs, NJ: Prentice-Hall Inc; 1994.

[10] Allemang RJ. The modal assurance criterion: twenty years of use and abuse. Sound and Vibration. 2003;37(8):14–23.

[11] Bayraktar A, Birinci F, Altunışık AC, Türker T, Sevim B. Finite element model updating of Senyuva historical arch bridge using ambient vibration tests. The Baltic Journal of Road and Bridge Engineering. 2009;4(4):177–185. DOI: 10.3846/1822-427X.2009.4.177-185.

[12] Sevim B, Bayraktar A, Altunışık AC, Atamtürktür S, Birinci F. Assessment of nonlinear seismic performance of a restored historical arch bridge using ambient vibrations. Nonlinear Dynamics. 2011;63(4);755–770. DOI: 10.1007/s11071-010-9835-y.

[13] PULSE. Analyzers and Solutions. Release 11.2. Bruel and Kjaer; Sound and Vibration Measurement A/S; Denmark; 2006.

[14] OMA. Operational Modal Analysis. Release 4.0. Bruel and Kjaer; Structural Vibration Solutions A/S; Denmark; 2006.

[15] Brownjohn JMW, Magalhaes F, Caetano E, Cunha A. Ambient vibration re-testing and operational modal analysis of the Humber Bridge. Engineering Structures. 2010;32:2003–2018. DOI: 10.1016/j.engstruct.2010.02.034.

[16] Bayraktar A, Altunışık AC, Sevim B, Türker T. Earthquake behaviour of kömürhan highway bridge using validated finite element model. Journal of Testing and Evaluation. 2010;38(4):467–481. DOI: 10.1520/JTE102102.

[17] Bayraktar A, Altunışık AC, Sevim B, Türker T. Finite element model updating of Kömürhan highway bridge based on experimental measurements. Journal of Smart Structures and Systems. 2010;6(4):373–388. DOI: 10.12989/sss.2010.6.4.373.

[18] Bayraktar A, Altunişik AC, Türker T. Structural condition assessment of Birecik Highway Bridge using operational modal analysis. International Journal of Civil Engineering. 2016;13(5):1–30.

[19] Ateş Ş, Constantinou MC. Example of application of response history analysis for seismically isolated curved bridges on drilled shaft with springs representing soil. Soil Dynamics and Earthquake Engineering. 2011;31(3):334–350. DOI: 10.1016/j.soildyn.2010.09.002.

[20] Altunışık AC, Bayraktar A, Sevim B. Output-only system identification of post tensioned segmental concrete highway bridges. Journal of Bridge Engineering. 2011;16(2):259–266. DOI: 10.1061/(ASCE)BE.1943-5592.0000150.

[21] Bayraktar A, Altunışık AC, Sevim B, Türker T. Modal testing, finite element model updating and dynamic analysis of an arch type steel footbridge. Journal of Performance of Constructed Facilities. 2009;23(2):81–89. DOI: 10.1061/(ASCE)0887-3828.

[22] Bayraktar A, Altunışık AC, Sevim B, Türker T. Ambient vibration tests of a steel footbridge. Journal of Nondestructive Evaluation. 2010;29(1):14–24. DOI: 10.1007/s10921-009-0061-9.

[23] Altunışık AC, Bayraktar A, Sevim B, Özdemir H. Experimental and analytical system identification of Eynel arch type steel highway bridge. Journal of Constructional Steel Research. 2011;67(12):1912–1921.

[24] Bayraktar A, Altunışık AC, Türker T. Structural health assessment and restoration procedure of an old riveted steel arch bridge. Soil Dynamics and Earthquake Engineering. 2016; 83:148–161. DOI: 10.1016/j.soildyn.2016.01.012.

[25] SAP2000. Integrated Finite Element Analysis and Design of Structures. Berkeley, CA, USA: Computers and Structures Inc.; 1998.

[26] ANSYS. Swanson Analysis System. USA; 2003.

[27] Sevim B, Atamturktur S, Altunışık AC, Bayraktar A. Ambient vibration testing and seismic behavior of historical arch bridges under near and far fault ground motions. Bulletin of Earthquake Engineering. 2016;14:241–259. DOI: 10.1007/s10518-015-9810-6.